Faces of Science

by V. V. NALIMOV

Edited by ROBERT G. COLODNY

Illustrations by MIKHAIL ZLATKOVSKY

ISi PRESS™

PHILADELPHIA

Published by

iSi PRESS™ A Subsidiary of the
Institute for Scientific Information®
3501 Market St., University City Science Center, Philadelphia, PA 19104 U.S.A.

© 1981 ISI Press

Library of Congress Cataloging in Publication Data

Nalimov, V. V. (Vasilii Vasil'evich), 1910–
  Faces of Science.

Translation from the Russian.

  Includes bibliography and index.
  1. Science—Philosophy. 2. Science—Methodology.
3. Mathematical linguistics. I. Colodny, Robert Garland. II. Title.
Q175.N225   501   81–6654
ISBN 0–89495–010–X   AACR2

Printed in the United States of America

Contents

Foreword

But as for certain truth, no man has known it,
Nor will he know it; neither of the gods,
Nor yet of the things of which I speak.
And even if he were by chance to utter
The final truth, he would himself not know it;
For all is but a woven web of guesses.

XENOPHANES OF KOLOPHON

An immense impulse was now given to science and it seemed as if the genius of mankind, long pent up, had at length rushed eagerly upon Nature, and commenced, with one accord, the great work of turning up her hitherto unbroken soil, and exposing her treasures so long concealed. A general sense now prevailed of the poverty and insufficiency of existing knowledge in *matter of fact*; and, as information flowed fast in, an era of excitement and wonder commenced to which the annals of mankind had furnished nothing similar. It seemed, too, as if Nature herself seconded the impulse; and while she supplied new and extraordinary aids to those senses which were hence forth to be exercised in her investigation — while the telescope and the microscope laid open *the infinite* in both directions — as if to call attention to her wonders, and signalise the epoch, she displayed the rarest, the most splendid and mysterious, of all astronomical phenomena, the appearance and subsequent total extinction of a new and brilliant fixed star twice within the life time of Galileo himself. [J. F. W. Herschel, The Cabinet of Natural Philosophy (Philadelphia, 1831), Secs. 106–107]

* * * * *

Our age is possessed by a strong urge towards the criticism of traditional customs and opinions. A new spirit is arising which is unwilling to accept anything on authority, which does not so much per-

vii

mit as demand independent, rational thought on every subject, and which refrains from hampering any attack based upon such thought, even though it be directed against things which formerly were considered to be as sacrosanct as you please. In my opinion this spirit is the common cause underlying the crisis of every science today. Its results can only be advantageous: no scientific structure falls entirely into ruin: what is worth preserving preserves itself and requires no protection. (Erwin Schrödinger, *Science and the Human Temperament*)

The rational investigation of the Cosmos is the last great enterprise begun by *homo sapiens*. If we start with the priest–astronomers of Sumer, we look back on only forty centuries, a mere tick on the geological clock. Great art is much older, and so are certain metaphysical and religious systems. Homer and the Psalmists precede Aristotle and Archimedes by five centuries.

If we judge the procedures, methods, and organization of modern science as defining this enterprise *for us*, then only ten generations separate us from the age of Galileo and Newton. And it is only in this generation, after immense upheavals and reconstructions in all of the scientific disciplines, that a mature, critical historical consciousness permits scholars to survey the totality of the scientific edifice, to search out the hidden character of the dynamism of science, and to complete preliminary surveys of the multiple interconnections of science with all other elements of culture and society.

This is an ongoing global enterprise—one in which the clash of contending schools is, fortunately, conducted in the muted footnotes of learned journals. And now there have begun to emerge the broad outlines of a consensus concerning where the problems are located in our effort to understand the essential parameters that have controlled the evolution of scientific knowledge and what are the dangers that threaten the continued existence of the scientific communities as well as that wider social world in which they are embedded. It is for these reasons that the probing essays of V. V. Nalimov are most welcome and enlightening. As their content demonstrates, these are not exotica from a remote and mysterious Tartary, but careful explorations of a familiar terrain: the world of Karl Popper, Thomas Kuhn, Kurt Gödel, Ludwig Wittgenstein, Niels Bohr, Norbert Wiener, *inter alia*. The author, a well-known mathematician and cyberneticist, is a member of the Scientific Council of Cybernetics and director of the Laboratory of Mathematical Theory of Experiment of Moscow State University. He came to this position after years of experience in physics and metallurgy, and extensive studies of the role of information flow between and through scientific communities. These studies had led him quite naturally to adopt a cybernetic view of the evolution of scientific concepts and to consider in depth the nature and function of the many languages of science.[1] As an expert in theoretical and applied statistics, Nalimov is in an extremely favored position to explore the most important shift in modern science, i.e., the transition from a world described in the deterministic mode to one analyzed by the calculus of probability. Nalimov is not merely concerned here with, say, the development of statistical mechanics and the description of the foundations of the world according to the mandates of physics; he turns this mode of description upon the very processes of scientific discovery and communication. Thus, in his vision, one can perceive a more intimate, human link connecting three elements: the inquirer, his language, and the external world that is under human scrutiny.

Perhaps the heart of the matter is indicated by the title of Chapter 2: "Scientific Creativity as a Manifestion of Intellectual Rebellion: A Bayesian Approach to the Problem."

As both a teacher and a researcher, Nalimov shares with us some of the problems of conveying to students and experimentalists the limitations to our knowledge imposed by a probabilistic ontology. The old world of Laplacean certainty is dead and beyond recall. The challenge to modern scientists and other humans is to operate rationally within an area of irreducible uncertainty.

There is no doubt that the conquests of the human mind have always aroused a sense of mystery in those who reflect deeply upon the history of our species' intellectual pilgrimage. The critical intellect of classical Greece was aware of this as well as the hubris that might accompany it (see Xenophanes, above). Aeschylus in *Prometheus Bound* was driven to ascribe the invention of science and the practical arts to the intervention of a rebellious Prometheus.

> Of rising stars and setting I unveiled.
> I taught them Number, first of Science;
> I framed the written symbols into speech,
> Art all recording, Mother of the Muse . . .

Nalimov searches out a different heuristic model. He examines the scientific community as a metaphorical organism, set within the biosphere. This is carried by the title of Chapter 7: "Science and the Biosphere: An Attempt at a Comparative Study of the Two Systems." These, of course, have as empirical links information flow, evolution and mutations, and ecological equilibrium. As far as this writer knows,

---

[1] The studies were first published in the USSR with the title *Probabilistic Model of Language*. The English version appeared as *In the Labyrinths of Language: A Mathematician's Journey* (ISI Press, Philadelphia, 1981).

this is a unique foray into the unknown. It clarifies the importance of the scientific community as a self-regulating and self-correcting system. It underlines the importance of the network of scientific information flows — a network which is coeval with the appearance of the first scientific journals and the first national academies of science in the seventeenth century. It pinpoints the danger of the breakdown of the system, either through political malevolence or through information overload and the choking of the circuits. Lastly, I would call attention to Nalimov's deep concern with the relations, often strained, between the sciences and the older humanistic traditions. Nalimov decries the tendency to isolate the scientific and the humanistic components of contemporary culture, and his brief critical analysis of this problem in the USSR will be useful for American academicians. For them, the problem posed by C. P. Snow eventuated in some turmoil at the higher levels of the academic establishment. This quickly subsided, leaving some rubble in the form of curriculum study and reform committees.

Nalimov's conclusions are similar to those of our own I. I. Rabi:

> The Universe is not given to us in the form of a map or guide. It is made by human minds and imaginations out of slight hints which come from acute observations and from the profound strategems of experiments . . . Science and the humanities are not the same thing; the subject matter is different and the spirit of tradition is different. Our problem in the search for Wisdom in a contemporary world is to blend these two traditions in the minds of individual men and women. [*Science The Center of Culture* (The World Publishing Co., New York, 1970), p. 33]

Towards this ancient and universal goal, Nalimov's *Faces of Science* is a valued and beautiful contribution.

ROBERT G. COLODNY
University of Pittsburgh

Preface

This book is the author's unrestrained contemplation of what science is. But these are not the thoughts of a historian or philosopher — they belong to a person who has almost all his life been a common soldier of science. This is almost an autobiography. I entered science as a youth: in 1931 at the age of 21, I started working at a laboratory of high vacuum technology and participated in research on the photoelectric effect. Much later, I worked for many years in the laboratories of metallurgical plants and geological survey boards. In 1957 I returned to Moscow and for three years worked with scientific literature at the All-Union Institute of Scientific Information (VINITI) of the Academy of Sciences; then I worked at a large metallurgical institute, where I organized the first group on mathematical methods of research. Finally, for more than a decade I have been a professor of statistics at Moscow State University, where I first was the head of a section and then of the laboratory of mathematical theory of experiment.

My early interest was in concrete research results: that was what I was paid for. It was often necessary to make very risky decisions, especially in the hard war years. One had, on behalf of science, to produce technological "wonders" which seemed to contradict science. However, everything came out in the long run. During this period, I was always bothered by fundamental problems as well. What is a good experiment? What is a good theory? Often, I used to make decisions under great uncertainty, and I was invariably attracted by the language of probabilistic ideas. Later the construction of probabilistic models and mathematical methods of experimental design became my profession. My interests broadened little by little until at last I got very much involved in the problems of the nature of science and of its penetration into social life.

By and by, I came back to the problems of *Weltanschauung* which had aroused my keen interest as a youth. I was one of the two authors of the book *Naukometriya (Science-of-Science)*, which was published in 1969 (Nauka, Moscow). There we tried to comprehend the development of science proceeding from the statistical analysis of data extracted directly from scientific publications. *Veroyatnostnaya Model' Yaszyka*[1] *(Probabilistic Model of Language)*, published in 1974 (Nauka, Moscow), was an attempt to comprehend from the probabilistic point of view the entire diversity of using a *word*, both in our culture and in the cultures of the past. This made possible a passage to a probabilistic comprehension of human *consciousness*. These problems are considered in detail in an almost-completed collection of papers, *The Past in the Present*,[2] which is a natural continuation of *Probabilistic Model of Language*. The present collection of papers, *Faces of Science*, seems to have emerged as a self-sustained book without any additional efforts—all the papers in it have been published previously. All these books, and especially the three later ones, are written in a highly similar manner. They are an attempt to discuss the central problem of today—the problem of *man* in the contemporary world.

Man is revealed through his vision of the world. *Science*, in the dialectical opposition of the logical versus the illogical, reflects *human nature* rather than the nature of that world described by man. Therefore, the study of the nature of science itself is primarily a way of understanding man.

I believe that it would be difficult to construct an all-embracing theory of science. Such a theory would unavoidably be incomplete and useless. But a paradoxical formulation of problems may help us to perceive new features in familiar and well-known phenomena.

It would be tragic if someone were to think that I am striving to create a school or a sect. My aim is not to convince or to argue. I am merely speaking of what I have thought over on the basis of my own experience and of other people's contemplations. Every honest and earnest person goes his own way, but he invariably takes into account other people's experience. I hope that my experience will be of some help to others and that the dialogue will continue (at least by correspondence). This may be the way to create a chain of sequential links. (And probably our entire culture is a net of such chains.)

Faces of Science is a collection of thematically related papers. Almost all of them have been published in Soviet scientific journals, and some of

---

[1] An English version of this book was published in 1981 under the title *In the Labyrinths of Language: A Mathematician's Journey* (ISI Press, Philadelphia).

[2] One of the chapters in this collection, namely, "Language and Thinking: Discontinuity vs. Continuity" was published in 1978 by Tbilisi University Press and appears in English translation as Chapter 6 of *In the Labyrinths of Language*.

Science models are rather a simulation of human consciousness than the reality of the universe.

them, in foreign ones. Some have appeared in collections of papers or in popular scientific magazines with a large circulation, but in the latter case they were always adapted for the general public, and the result was something like a *Decameron* for children. Some of the papers have been published as preprints, and one of them was published in only 200 copies.

Each of the papers reflects a different aspect of science. Some of the papers are of a logico–methodological nature; others are devoted to the problems of the science-of-science. The whole book has a cybernetic character: science is viewed as a large self-organized information system.

In an unconstrained dialogue with the well-known British philosopher Sir Karl Popper, I indicated that quite unexpected conclusions can be drawn from Popper's constructions. Using the system of Bayesian concepts, I proceeded to strengthen Kuhn's statements on the role of paradigm in science; scientific creativity is regarded as a manifestation of intellectual rebellion.

In analyzing the logical foundations of the mathematization of knowledge, the structures of pure and applied mathematics are contrasted. The development of a probabilistic language for describing the world is considered historically, and the emerging problems of logical foundations are discussed again. Bohr's principle of complementarity is regarded as a

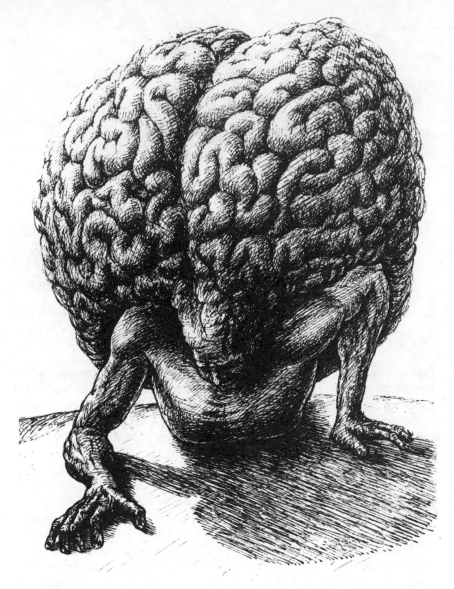

One of the Faces of Science

rejection of the law of the excluded middle. The penetration of the humanities into other fields which has begun lately is discussed. Science as a self-organized information system is compared with the biosphere. Certain aspects of a possible approach to solving eschatological problems are outlined (it is in this way that I perceive the problem of the ecological crisis), an almost Zen analysis of such notions as "goal" and "victory" is given, and a probabilistic model of social behavior is suggested. The struggle with complexity in scientific descriptions of the world is illuminated; difficulties in constructing theoretical biology are formally analyzed.

As to the science-of-science, I touch upon how scientists of different countries and of different cities in the same country are provided with information. The data of statistical analysis illustrate the way science is evolving from the physico–chemical into a predominantly biological mode.

All these papers collected under one cover look different than they did when published separately. In the slightly dissipating fog, we come to distinguish the contours of an absurdly complicated, whimsical, and cumbersome Edifice of Science which, far from being completed, already undergoes perpetual rebuilding and is partly demolished in several places.

Chapter 1

The Structure of Science

Logic of Accepting Hypotheses[1]

*Our knowledge can be only finite, while our ignorance
must necessarily be infinite.*

KARL POPPER

*Not the aspiration for discovering indubitable truths
should be made the principle, but the ability to find the
dubitable in what traditionally was considered
indubitable.*

A. LUBISHEV

Introduction

Nowadays, Western philosophers pay great attention to the philosophy of science. This is a new branch created in epistemology, constructed, like the natural sciences, on the basis of the analysis of real experience. In this discipline an attempt is made to estimate the human potential for cognition of the real world based on the analysis of the way science is arranged and scientific knowledge develops. The history of science becomes the foundation on which we base judgments on the nature of our knowledge. Two brilliant examples of such an approach are the book by Kuhn (1970a) and an extensive paper by Lakatos (1970).

I shall not try to present here in any adequate way the multiplicity of

---

[1] A summary of this chapter was published in Russian in the collection of abstracts of the reports at the symposium "Istoriya Nauki i Naukovedenie" ("History of Science and Sciencemetrics"), Riga, 1975, in the journal *Khimiya i Zhizn (Chemistry and Life)*, no. 1, 1978. An almost complete text was published in English and in French in the journal *Diogène*, no. 100, 1977, and in Polish in the journal *Zagadniena Naukoznawstwa*, 1978. This chapter was translated by A. V. Yarkho.

ideas connected with the analysis of the structure of science. I shall confine myself to exposing my own views on these problems. I am not a philosopher and still less an expert in the history of science. My views are based upon experience obtained through regular contacts with representatives of numerous fields of knowledge, since, being a statistician, I have had to participate professionally in discussing various research problems. Thus, my examples will relate not to the history of science but to its contemporary state.

Problems in Science

Were it not for a fear of certain schematization, I could point to three basic and obvious structural constituents in science: *problems*, i.e., questions to be answered; *hypotheses* whereby these questions are answered; and, finally, *means* which allow these hypotheses to be accepted or rejected.

We seem not always to understand completely the important role played by well-formulated, accepted, and permitted questions in our intellectual activity. Langer (1951), developing the idea previously stated by Cohen (1929), even believes that the development of every culture can be characterized by a set of questions, some of which are permitted and formulated and others of which are forbidden. This already envisages the limited possibility of rational responses, or, as Langer puts it: a question is an equivocal sentence whose determinant is its response.

The difference between cultures is first of all the difference in the questions permitted. This statement is best illustrated by the version of the dialogue between Christ and Pilate given in the Gospel according to St. John. Christ says, at the interrogation by Pilate, "I came into the world for this, that I should bear witness to the truth." Then Pilate asks his next question, "What is the truth?" The question remains unanswered (although elsewhere Christ says that He is the truth).

Pilate, being a man of Hellenistic culture, has to begin by discussing the question of what truth is. Christ is a representative of another culture in which the question is forbidden. Christianity, at least in its early period, dealt with other questions, such as those treating good and evil, or the question of retribution, but they all gained the right to exist as long as the question "What is truth?" was forbidden.

The state of science at every stage of its development is revealed by a set of permissible questions. It is easy to give examples of absolutely forbidden questions of the kind "Whence came Ohm's law and what does it exist for?" or even stronger ones such as "How, why, and with what purpose were the laws of nature formed?" and "What is the goal of the world's existence?"

Everybody who has extensive teaching experience knows how vexing are the students who ask questions forbidden by science. Usually the teacher gets irritated, and his only resort is to ridicule the student in some illicit way.

Sometimes science transcends itself and answers questions yet unformulated. In this case the answers prove untimely. Mendel answered a question yet unformulated in his epoch and for this reason his work was not acknowledged for a long time. At the same time, many clearly formulated questions remain unacknowledged for long periods. This can be illustrated by the case of Malthus. His positive statements have proved wrong: as has become obvious, all the troubles arise because production for consumption grows exponentially, resulting in the menace of exhausting the earth's resources. But Malthus formulated a question which had long been considered unlawful by many scholars.

Retrospectively, it seems that science can be regarded as a series of answers to a number of profound questions. This process of development, like any other, requires limitations, and these are imposed by the fence of forbidden questions. However, the same is probably true of theologies: they develop while questions are allowed in the system of their ideas. Again, retrospectively, we can agree with Kuhn's remark (1970a) that astrology stopped in its development not merely as a result of its neglecting to test predictions but primarily as a result of the absence of questions within it.

Profound questions, either explicit or implicit, are a feature which seems to characterize each ideologically developing system. However, this is not yet the demarcation line between science and non-science. This is a necessary, but not sufficient, characteristic.

It is interesting that, in analyzing the science of today, it is not at all easy to single out those branches in which well-formulated questions exist. Start looking through monographs and reviews, and you will find the report of what has been achieved rather than formulations of questions. This is why it is so difficult to separate the science of yesterday from the science of tomorrow.

I shall remind the reader that Hilbert, an outstanding mathematician of the recent past, in 1900, at the International Congress of Mathematics in Paris, formulated his famous 23 problems of mathematics. Among them was a problem the answer to which became Gödel's famous proof. This is one of the most brilliant achievements of the human intellect, but it was accomplished only in 1931. Hilbert's initiative in formulating the core problems was not supported.

One branch of mathematics is mathematical statistics. An interesting conference on the future of statistics was held in 1968, at the University of Wisconsin (Watts, 1968). The reports presented revealed the fact that nobody could formulate the central problems of mathematical statistics. The future of statistics is in the communication of statisticians with researchers in other branches of learning. It is honestly recognized that the statistics of today is a purely methodological science, and its future will depend on the way the principal problems are formulated in other fields of knowledge for which mathematical statistics plays an auxiliary role.

Today questionnaires are in wide usage, including those addressed to scientists. However, one crucial question is never asked: What is the future of your field of knowledge? Once my co-workers tried to put the question to specialists in chemical kinetics, but they did not receive an intelligible answer. Probably we just do not know how to phrase the question properly.

I am usually impressed by the conferences, discussions, reports, and publications where new questions are formulated. But often I have participated in scientific conferences where I hear answers to unformulated questions. Sometimes these are just specifications or slight developments of something already done, sometimes an example to support something already known, or, finally, mere comments upon something already said.

At present, the attenuation of the previously existing exponential growth of allocations for science raises the question of redistributing funds among various fields of knowledge. Here one needs a criterion. Perhaps the ability of scientists to formulate explicitly the central problems of their fields of knowledge in a meaningful and original way will become such a criterion.

How Hypotheses Are Formulated

It is common knowledge that the success of any research depends primarily upon the way the problem, i.e., the initial hypothesis, is formulated. But we know very little of how fruitful hypotheses are formulated. We cannot suggest any model, even a very weak one. In comparing human abilities with those of a computer, Kendall (1966) noted that one of the cardinal differences lies in the fact that a man observing new phenomena can formulate new fruitful hypotheses; we have not so far managed to teach a computer to do this. Inductive logic proves *unyielding to expression in algorithms*. What is it then? Here it is of importance to know that the models we are so accustomed to in science can be obtained only from premises and not immediately from observational results. This well-known principle of formal logic is constantly forgotten, and one often comes across the statement that a hypothesis (or a mathematical model) is inferred from observational results. New observations can certainly serve as an impetus to the formulation of new postulates, but the mechanism of this impetus is unknown. It must be described not in terms of logic but in terms of psychology. But discussing the problem in terms of psychology does not enable us to state anything definite either. In practice, researchers with profound erudition and critical wit often prove quite impotent in formulating novel hypotheses. Like Karl Popper (1963), we must acknowledge that the first peculiarity or, if you like, the first paradox in the development of science is the creative constituent. The process of formulating novel hypotheses does not possess any traits unique to science. In any case, it cannot be distinguished from myth creating.

Popper's refusal to consider important how hypotheses are formulated gives his concept the serious flavor which impresses scholars. As I have already said, in science there exists an implicit but universally acknowledged ban: do not ask questions which cannot be answered in the system of accepted structures. Some questions turn out to be so constructed that a response to them can be given only in mythological structures. Popper starts his theory by removing the question of scientific epistemology, the answer to which would have diverted scientifically minded readers from him. "Whereof one cannot speak, thereof one must be silent," says Wittgenstein in his *Tractatus*.

Popper's statement that scientific hypotheses are just conjectures also has a great methodological value, since, if it is so, then at the moment of formulating a hypothesis one need not trouble about its foundation; it is more important, according to Russell (1961), to believe in it, supposing that it proceeds from a certain intuitive, i.e., merely inexplicable, motive. Serious reasons on which to base it can be obtained only during the subsequent theoretical or experimental development. It is only unfortunate that already at the first stage, at the moment of formulating a hypothesis, a scientist has to say something to account for what he cannot as yet account for. (The ideas expressed in this paragraph were prompted by a discussion with Yu. A. Shreider.)

Nonetheless, the right freely to formulate hypotheses should not allow schizophrenic ideas to penetrate into science; these, however, are easy to discover.

How Hypotheses Are Accepted

Here we have to accept Popper's second paradox (1963, 1965): scientific hypotheses, in a broad problem formulation (this is our reservation), may be only *falsifiable* and in no way *verifiable*. This statement, though in a slightly different form, can be found in any serious handbook of mathematical statistics. Indeed, if we consider only one hypothesis (say, that the results of observations can be represented by a straight line) without opposing it to other hypotheses not formulated by us, the conclusions from the results of experimental investigation of the hypothesis' true nature (if it is not refuted by these results) may be recorded only as "Our hypothesis does not contradict our observational results," since the observed phenomena can also be consistent with other hypotheses, not formulated by us. The answer recorded like this looks very weak. We can strengthen it if we start considering not one but several rival hypotheses (for more detail about this, see Nalimov, 1971). Then the result of a well-designed experiment will enable us to choose one of the rival hypotheses (if they are formulated as antagonistic ones). Here we can already say that one of the hypotheses has stood the test and, consequently, proved verifiable. However, this is verification only for the narrow problem formulation—the hypothesis is verified only with respect to the considered set of clearly formulated rival hypotheses, under the assumption that there is a true one among them. Here again, we cannot speak of verification in the strict sense, as our specially designed experiments will prove consistent also with respect to some other hypotheses not included in the set in question.

Thus, in contrast to positivists, Popper believes that, strictly speaking, it is impossible to speak of verifiability of hypotheses. No matter how many factors support the hypothesis, we shall never be able to state that its acceptance is final. A scientific hypothesis is always open to further testing, and a single new inconsistent result is enough to reject it. This asymmetry in testability, when any number of supporting factors are not sufficient cause to accept the hypothesis and a single negative factor is sufficient for its refutation, is of immense methodological importance.

According to Popper, every truly scientific hypothesis should be formulated so that it could be tested by a crucial experiment. If the hypothesis stands the tests, i.e., is not falsified or refuted by a crucial experiment, it is provisionally accepted but remains open to further testing. Vaguely formulated hypotheses in the frame of which any experimental result can be explained lose scientific character if they are regarded from this point of view.

Popper states that the progress of the natural sciences consists in the very fact that their structures always remain open to further testing. In contrast to positivists, Popper calls himself a "negativist."

Popper's concepts started a broad discussion in the West. The above-mentioned article by Lakatos (1970) cautioning against naive and dogmatic understanding of falsificationism proved especially interesting. Indeed, despite the amazing logical precision of the idea of falsificationism, we still have to acknowledge that in reality science does not develop according to such a simple scheme. If we want to test a hypothesis, we must first of all agree as to what will be a crucial experiment in this case.

Any real experiment gives rise to errors, and for this reason even if we test a quite elementary hypothesis—say, that of approximating observational results with a straight line—we must introduce the notion of significance level and formulate our answer roughly as follows: for the significance level we have chosen, the divergence between our model and observational results is not statistically significant. In choosing the level of significance, we introduce a system of conventions into our methodology. Further, we have to convince our potential opponents that our experimental results are free from systematic errors. For this purpose, a procedure of randomization is introduced: experiments are conducted so as to be randomized with respect to those variables which may change uncontrollably during the experiment. The choice of variables according to which the experiment is randomized is again determined by the level of our knowledge and that of our potential opponents. We choose a randomization scheme with the hope of convincing our opponents by means of it. Again, we have to recognize that we are relying on a system of conventions. Finally, if an experimental result is obviously inconsistent with a very serious hypothesis, it does not yet follow from this fact that the hypothesis will immediately be rejected. There are other possibilities: the experiment may be interpreted in a special way, or the hypothesis may be slightly modified. A negative experiment becomes crucial when a novel, powerful counterhypothesis is formed under its effect. Lakatos (1970) brilliantly illustrates this idea by the Michelson-Morley experiments. Citing Bernal, Lakatos says that the Michelson-Morley experimental data were the greatest negative result in the history of science. But, he goes on, what was it negative to? Everything was not that simple—the crucial role was played by novel ideas.

Lakatos also reminds us that the history of science provides us with numerous instances of accepting hypotheses. They are accepted when they turn out to predict new and interesting phenomena. True, it is probably more pertinent to speak here not of hypotheses but of programs for action. Further, Popper's mechanism of falsifiability is sure to start working.[2]

---

[2] However, in the history of science there are interesting illustrations of how the hypotheses falsified by the experimental results still were not rejected. The well-known biologist Jacques Monod describes the trap in which Darwin's theory was once caught (Monod, 1975). From Popper's point of view, Darwin's theory is a second-class theory, since it is impossible to subject it to a crucial experiment. But it once seemed that such crucial conditions actually existed. William Thomson, Darwin's contemporary, was among the few physicists who could calculate astrophysically, and he showed that, assuming the sun is composed of coal (one of the most calorific fuels known at that time), its energy would never suffice to warm the earth during the period necessary for the evolution of life. This was a purely experimental

Now a few words about Popper's demand that hypotheses be formulated so as to enable their falsification. In mathematical statistics we also always come across the demand to test investigated hypotheses. One of the problems of experimental design is precisely that of choosing experimental points from the region of values of independent variables where the greatest divergence in the process of discrimination of rival hypotheses is expected (Nalimov, 1971). However, it is not rational to demand that scientific hypotheses always be formulated so as to make them susceptible to a crucial experiment. Sometimes the logic of judgments makes us formulate hypotheses in a loose way from the viewpoint of their falsifiability. I shall illustrate this with an example.

In our study of the history of the development of science (Nalimov and Mul'chenko, 1969), exponential growth curves were used which are easy to falsify. However, it is logically difficult to support the correctness of presenting the growth of scientific papers by one exponent. It is more reasonable to think that different fields of knowledge and different countries enter the game at different moments and with different constants of growth, and if this is so, the curves of growth should be given by the sum of exponents. Such a function is very hard to test experimentally. Too broad a class of curves can be presented in this way. A logically precise hypothesis proves amorphous in testing. This is often unavoidable, but in any case it is good to be aware of it.

Popper's concept also proves insufficient to explain the way scientists behave when it is impossible to conduct a crucial experiment allowing a choice of one of the rival hypotheses. In this case, too, one of the hypotheses is still chosen on the basis of some outside reasons. Let us illustrate this with an example.

Consider the problem of growth of the number of scientific workers or publications with time. In our paper cited above, following Derek de Solla Price, we described this process by an exponential curve, or a logistic one, and, on a broad time scale, by the sum of exponential curves. The divergence of points from the curves was interpreted merely as a fluctuation, since it was conditioned by many factors that were difficult to take into account. However, another approach is possible: that of describing the growth of publications with an epidemics model (Nowakovskaya, 1972). The material at hand does not allow us to prefer one of these approaches, but a mental experiment—extrapolation into the far future—testifies to the advantage of the first approach. It is difficult to imagine that science might develop as epidemics do: it will not fall to zero value. Still another approach is possible. In one paper it was proposed

refutation of Darwin's theory, since the dimensions of the sun and the amount of heat necessary for the evolution of life are experimentally determined values. Darwin was depressed by these calculations and introduced rather ruinous corrections into the second edition of his book. However, his theory was not rejected. Monod remarks that at present we know the theory implicitly contains the concept of solar atomic energy, though nobody could have had such an idea at that time.

that the time scale be divided into a number of separate sections so that for each one its own law for the increase in the number of scientific workers could be described by one or another differential equation. Equation parameters can be chosen so that the divergence between the model and the experiment is negligibly small. The author managed to logically join them with each other, showing the complicating process of growth. Outwardly, everything looked so respectable that the author said he had inferred his models from observational results. However, the absurdity of this approach is quite evident: it is easily shown that local changes of the growth curve are due to purely random phenomena, such as insufficient accuracy of the statistics on scientific workers, modifications in the instructions for accounting, or changes in the funds allocated for scientific work due to temporary difficulties.

In some cases, we simply cannot test the essentials contained in our theory. I shall again use a historical example. In one of the problems of chemical kinetics, five rival hypotheses were suggested for describing the mechanism of the phenomenon. Specially conducted experiments (carried out in the easily achievable interval of varying the variables) showed that the first of these models obviously yielded the worst results and for this reason seemed to be unconditionally rejected as not having stood the test of falsification. However, in extrapolation, it yielded better results, and this is not hard to explain. As a matter of fact, we tested only the interpolatory power of the model, but in no way its capacity to reflect the real mechanism of the phenomenon. The experiment could not be organized so that it would yield information on all intermediate responses hypothetically contained by the model. But if the possibilities of our test are such, then why should we develop models in so complicated a manner? Is it not easier to confine ourselves to presenting results by quite simple interpolational formulas, say, by polynomial models, without trying to probe the mechanisms of phenomena? If we behave so, we shall deprive chemical researchers of scientific imagination. But what is the permissible degree of scientific imagination?

I believe that in science there exist *protective mechanisms* other than falsifiability which allow one to make a *stabilizing* selection, but here we already start borrowing terms from the theory of evolution.

Paradigm — a Protective Mechanism in Science

> *For everything its season, . . .*
> *a time to plant and a time to uproot*
> ECCLESIASTES

Kuhn's (1970*a*) greatest merit is that he has introduced the notion of paradigm as an intellectual field driving the development of a particular branch of knowledge in one strictly defined direction and protecting it from the destructive influences of other possible approaches. I shall not try to give here a strict definition of a paradigm—this is impossible. Masterman (1970) has calculated that, in his book *The Structure of Scientific Revolution*, Kuhn uses the term "paradigm" in no less than twenty-one various meanings! Kuhn's theory started a very interesting discussion. The collection of articles *Criticism and the Growth of Knowledge* (Lakatos and Musgrave, 1970) is in its major part devoted to it. Among the articles in this book are "Logic of Discoveries or Psychology of Research" and "Reflection on My Critics" by Kuhn, "Against 'Normal Science' " by T. W. W. Watkins, "Does the Distinction Between Normal and Revolutionary Science Hold Water" by S. L. Toulmin, "Normal Science, Scientific Revolution and the History of Science" by L. Pearce Williams, "Normal Science and Its Dangers" by Karl Popper, and "Consolation for Specialists" by P. K. Feyerabend. The titles speak for themselves.

I shall not repeat this discussion, but shall only present my own views on the subject. I emphasize that a paradigm is, in my understanding, a stabilizing selection, i.e., a protective mechanism shielding at a certain stage of its development some trend from pollution or from spreading into lateral infertile areas. At another stage, it may hamper the emergence of new trends. Every outstanding scholar begins his career fighting against some old paradigm and creating a new one, his own. This idea is illustrated below.

The statement by Norbert Wiener that 95 percent of the original work in mathematics is done by 5 percent of mathematicians, the remainder only playing a protective role in sheltering it from being polluted by insufficiently strict constructions, is widely known. But in what way is this done? The notion of proof itself cannot be strictly formalized, as follows from Gödel's proof (for details, see Kleene, 1952). A propos, some serious authors (e.g., Nagel and Newman, 1960) affirm that the corollary from Gödel's proof is the impossibility of creating an artificial intellect, since it must be built only on the basis of deductive logic. But it has never occurred to anybody to deny the possibility of constructing mathematics as a deductive science. It still remains unclear in what way mathematicians cope with the non-formalized elements in their science.

One of the relevant problems may be formulated as follows: "How can mathematical statistics exist in the frame of mathematics?" American statisticians boast that they have divorced themselves from pure mathematicians. That gave them the opportunity to organize their own departments in many universities, and these are quite independent from mathematics departments. They have their own journals, and they have their own notion of prestige. Statisticians orient themselves not so much to the strictness of constructions as to the significance of the experimental data

The Struggle Against the Paradigm

obtained as a result of applying these constructions. There is nothing of the kind in the Soviet Union. There, statistics is either an appendage to economics or a branch of pure mathematics. In the latter case it must be as rigorous as pure mathematics. And it is in this frame, with contempt for experimental research and the corresponding mode of thinking, that students are brought up in the disciplines of mathematical statistics and probability theory. There are no Soviet journals like *Biometrika, Bio-*

metrics, and *Technometrics*. If they were to appear, they would probably be prohibited as non-scientific. The journal *Probability Theory and Its Applications* is published in the Soviet Union, but it is very difficult to find any applications there. The manuscript of a collective monograph, *Analysis and Design of Experiments*, has remained unpublished since 1971; nobody can decide whether it meets the unwritten standards of rigor. Occasionally, nonrigorous foreign books such as *Applied Regression Analysis* by N. R. Draper and H. Smith (*Statistika*, 1973) or *Spectral Analysis of Economic Time Series* by C. Granger and M. Hatanaka (*Statistika*, 1972) do get translated into Russian. (By the way, one Soviet mathematician called the translation of the latter book into Russian a mistake.)

Indeed, is it possible to apply spectral theory to the description of real economic phenomena? All really observable time series in economy are essentially non-stationary; all realizations are too short-termed. They must always be stuck together somewhat arbitrarily. Different months have different numbers of days, etc.) However, much earlier, in 1934, *Manuel des Calculs de Laboratoire* by H. Vigneron was published in Russian. This work set forth the general impossibility of applying mathematical statistics in laboratory measurements. Indeed, how is it possible to speak, e.g., in analytical chemistry, of applying the Gaussian error law (which, furthermore, is called normal) if it assumes (though with small probability) the possibility of any great error. It is quite evident that a load of several kilograms will simply break the beam of analytical scales. Analogous argumentation used to be popular among biologists. The book by Vigneron was a desk companion for years running in many laboratories. At present, this type of argument seems strange to everybody. The paradigm has changed, at least in this respect.

As a matter of fact, real problems in applying mathematical statistics do exist. A statistician almost never has a chance to test his hypothesis that the basic premises underlying the methods are realistic. Sometimes this may be avoided by means of so-called robust (i.e., insensitive to initial premises) estimation, but then we have to give up any theory of optimal estimations and resort to simulation models for the choice of techniques. Further, we often have to face the situation that a set of inconsistent optimality criteria[3] may be suggested for solving a problem. In what way is one of them to be chosen? Sometimes one has to resort to techniques quite incorrect from a mathematical standpoint, e.g., to using divergent series in constructing an algorithm for forecasting non-stationary ran-

---

[3] Criteria may be regarded as axioms. They do not form a mathematical *structure*, which, according to Bourbaki, is the most essential thing in mathematics as a science. Instead of a structure of mutually consistent axioms, we have to deal here with a mosaic of statements. For greater detail, see Nalimov (1974*a*) and Chapter 3 of this book.

dom processes. Finally, the most important thing is that the language of mathematical statistics is not context-free. This means that the legitimacy of statements in this language depends not only on the relations of symbols but also on the context — on what the researcher says. The mathematical langauge is not closed here. In what way, then, must a statistician behave if he has graduated from a mathematical college? He relies upon a system of implicit but universally understood conventions forming the paradigm of the clan of statisticians; this is not acknowledged by pure mathematicians.

A paradigm of mathematical statistics has not been created in the Soviet Union, but a vacuum is always filled in one way or another. There began to appear papers written quite "freely," not controlled or restricted by anybody. Some of them contain real blunders: conditional and unconditional probabilities are confused or the derivative of the constant proves not to equal zero, and all this is camouflaged with dozens of pages where something is first differentiated and then integrated again with some mysterious purpose.

However, what happens more often is not operational blunders but arbitrariness of interpretation in passing from a mathematical language into that of the physical world. In one paper, by means of solely mathematical methods, it was proved that human thinking can have no more than seven levels of abstraction! In another paper, in a similarly serious way, the possibility was stated of proving mathematically that in any branch of knowledge 50 percent of publications relate to this branch and the rest to the adjacent ones. In a paper on forecasting, it was stated that every seventh point possesses a peculiar property: it always gets on the curve. I could cite similar examples endlessly.

In Western literature the problem of whether science should be regarded as a rational structure or as an irrational one is debated very acutely.

Popper is supposed to be a supporter of the first view, and Kuhn, of the second. Feyerabend, answering this question, says, "Yes and no." Yes, science should be regarded as irrational because there does not exist a unique and constant set of rules for decision making as to what a scientific judgment is. No, it is not irrational since every step is made on the basis of logical judgments. I would add that people, including scientists, cannot act without limiting themselves by a system of postulates, which probably are not always clearly formulated. The mechanism of their formation is unknown: for its explanation we must resort to psychology rather than logic. At the same time, people are afraid of boredom: the moment a situation becomes dull and invariant there arises protest, a fight against a previously established paradigm. This is a psychological fact. In science this is a peculiar *way of fighting against Gödel's difficulty*. The notion of a paradigm may prove useful in solving problems re-

One View of Paradigms in Science

lated to the search for optimal ways of scientific progress. Organizationally, science must be structured so as to enable a moving equilibrium
between stability and changeability. In this respect the Soviet scientific
community, especially the structure of the Academy of Sciences, presupposes extreme stability. Election of academicians for their lifetime, their
right to select new members from the people coming into line with their
views, and their ability to influence significantly the development of
science must have been very favorable at a certain stage since this promoted concentration in strictly given directions and cut off everything
hampering this progress. But later, the right was given by the Presidium
of the Academy of Sciences to the journals to deposit (without publishing) certain manuscripts. Is not this one of the ways to freeze existing
paradigms? Everything essentially new, precisely because of its novelty,
will be deposited as not interesting for a broad circle of readers.[4]

---

[4] Paradigmatic pressure naturally manifests itself primarily in the reviewing of manuscripts before their
publication. Recently, the first investigation appeared which compared the reviewing of manuscripts with
their citation rate after publication (Gordon, 1977). The author concluded that ". . . expert estimates are
significantly uncorrelated with the number of citations after the publication of a paper. Moreover, the
results of analysis of the most frequently cited chemical journals are far from being flattering to the re-
viewers. There is even a negative correlation between expert estimates of the papers and their subsequent
rate of citations." This is also true of exact sciences. Is this not sufficient grounds for becoming more
tolerant in science?

Another View of Paradigms in Science

Mathematization of Knowledge as an Example of Constructing a New Paradigm

Attempts are often made to regard the mathematization of knowledge as a process directed at turning mathematized knowledge into a calculus. But, strictly speaking, we should have dealt, as Hutten (1956) has pointed out, with:

(1) strictly defined terminology
(2) an internally consistent system of axioms
(3) rules of inference
(4) rules of interpreting abstract symbols in experimental terms.

Even modern physics, Hutten goes on, does not satisfy these requirements. Physical terms do not, as a rule, yield to strict definition; only a few branches of thermodynamics are axiomatized in modern physics, and precisely formulated rules for interpretation are lacking.

It seems more rational to say that the mathematization of knowledge is merely reduced to using mathematics as a language. Many people believe that we disparage the role of mathematics in stating that, for non-mathematicians, mathematization of knowledge is but resorting to a novel language. As a matter of fact, the state of things is quite different: in speaking of mathematics as a language we attach to it extremely great importance. Language influences people's thinking, sometimes changing their *Weltanschauung*. This was clearly understood by some linguists and logicians. In Wittgenstein's famous *Tractatus* we read: "The limits of my language mean the limit of my world." Much earlier, Humboldt, speaking of differences among languages, stated, "their difference is not only difference in sounds and symbols but in *Weltanschauung*." The Russian linguist Potebnya wrote, "Language is not a means to express the ready idea but to create it . . . It is not a reflection of the formed outlook but the activity forming it."

These statements are brilliantly supported by what an experimenter feels when he perceives the probabilistic language of mathematical statistics. His thinking becomes probabilistic; he begins to view the surrounding world from a new angle and makes decisions in a new way. First of all, he begins to understand that he must not be afraid of chance. Moreover, he is convinced that experimental conditions should be *randomized*. The behavior of all uncontrollable variables must be made random; otherwise, experimental results will prove biased, distorted with systematic errors. Such a randomized scheme is not easy to realize, but statisticians can randomize within limits. All regimes cannot be tested in one day. However, a biologist working with us recently said to me, "We are carrying out non-randomized experiments again — this is simpler."

"And what about results?" I asked him.

"I am quite sure of the results."

"What do you need experiments for?"

"To convince others."

For this biologist, probabilistic language remained foreign and his outlook has not changed. The outlook of the people he is going to convince has not changed either. The paper will be published, but it will seem very unconvincing to biologists influenced by the ideas of modern statistics. Scientists working in the same field may be under the influence of different paradigms.

Let us consider another example showing how, under the influence of a probabilistically created paradigm, the experimental solution of multivariate problems changes.

If earlier a student was taught that in a well-designed multifactorial experiment every factor must be varied separately from the position of probabilistic judgments, it becomes clear that this is the worst possible course of action. Indeed, assume that our task is to build a polynomial of the first degree, varying the variables at only two levels. If the levels of variation are coded with numbers $+1$ and -1, then a unifactorial experiment allows us to estimate the regression coefficient b_i with variance $\sigma^2\{b_i\} = \sigma^2\{y\}/2$, where $\sigma^2\{y\}$ is the experimental error variance. Another experimental design where all the variables will be varied according to a special scheme, namely, the Hadamard matrix, allows us to estimate regression coefficients with variance $\sigma^2\{b_i\} = \sigma^2\{y\}/(k + 1)$, where k is the number of independent variables. Imagine that you deal with fifteen independent variables; then we shall have a gain of eight, though the total number of observations N will be $N = k + 1$ for both unifactorial and multifactorial experiments. This fantastic result is achieved fairly simply. This possibility has long remained unnoticed only because the debate was not held in a probabilistic language. Few experimenters understand this at present.

The number of examples can be increased. If, say, a researcher has to estimate the number of experiments in quality-control problems, probabilistic language will immediately force him to formulate the conception of first- and second-type errors, i.e., to give the probability of accepting an unfit production and rejecting a fit one; this makes him replace the point boundary between fit and unfit by an interval boundary. When we have to make decisions after a new experiment, the probabilistic language makes the researcher take into consideration his previous experience, also probabilistically given by the prior distribution function. Here the Bayesian theorem will be used for decision making. The solution will be obtained in terms of many-valued or, if you like, even continuous logic, not in the traditional bi-semantic one.

This shows that, having begun to use probabilistic language, we also begin to *think* in a new way. The same may be said of the impact of other mathematical branches on the researcher's thinking as well. It is often said that the advantage of mathematics is its capacity to pass from qualitative propositions to quantitative ones. As a matter of fact, mathematical language, when applied to other fields of knowledge, does something more significant: it increases the level of abstractness of judgments. This idea is explicitly developed by Hutten (1967):

> The main import of mathematics is that it provides a universal system of symbols rather than merely a means for quantitative judgement . . . The use of this symbolism allows us to widen the horizon of our knowledge beyond immediate experience. Science is the "abstract" representation of reality. We build up science by constructing more and more abstract theories.

A higher level of abstractness of our scientific theories means another change of a paradigm in science.

The following question seems quite proper here: Where will the change in the scientist's outlook lead — to creating new research programs of great heuristic power or to constructing only locally correct models explaining and even predicting something but not opening new vistas in scientific progress? This question is not easy to answer. Broad mathematization of knowledge has only begun recently, but already we can give both examples and counterexamples. In linguistics, the abstract mathematical conception of context-free languages generated by Chomsky and especially that of transformational grammar has created a new research problem taken up by broad circles of linguists and productive of interesting philosophical generalizations. Another example is the fertilizing effect of mathematics upon the development of genetics. However, counterexamples may be given, too: substantial application of mathematics in ecology, physiology, epidemiology, sociology, and psychology seems so far to have resulted only in creating locally correct models. It may be that many will disagree with me, but this is an interesting matter to debate.

Finally, I would like to emphasize once more that the process of falsification itself is determined by the paradigm accepted at the moment. This statement can best be illustrated by what is happening before our eyes. Probabilistic thinking and the ensuing statistical methods of experimental design and of testing hypotheses form conditions for falsification which introduce such notions as randomization of experiment, distribution functions and confidence limits, model adequacy (in the statistical sense), first- and second-type errors, use of prior information, and the expression of experimental results as a posterior distribution function. But how are we to make use of this store of techniques when various probabilistic approaches are equally legitimate? Strictly speaking, there are no answers to these questions. This is our paradigm—the paradigm some scientists have accepted and others have not. And here we must formulate an important addition to Kuhn's conceptions: different groups of scientists can simultaneously work and coexist under the influence of different paradigms.

It is important to note that nowadays, when science can easily generate new powerful means, the problem of a paradigm controlling the rules for testing hypotheses is not of an abstract character. This may well be illustrated by the notorious story of thalidomide. In West Germany alone, this drug resulted in the birth of 6,000 defective babies. At the moment of its production, the drug was said to be nontoxic. This hypothesis had stood the test: no rats died, even from the strongest dose. Later, this hypothesis was transformed into the statement that the drug was harmless for humans. This more powerful hypothesis was not tested; in particular, the medicine's effect upon embryos was not tested. Nowadays, too, when new drugs or vaccines are produced, adequate testing of them can never be guaranteed, since, as a rule, nobody can test their possible effect after a long period of time. The thalidomide tragedy was aggravated by the fact that, when the first alarm signals began to appear, the firm producing the drug ignored them, alluding to the "statistical character of contraindication," the "dubitable character of causal connection," and the "absence of direct proofs." The story of thalidomide is well described in an article by Bross (1964), who believes that an important role was played by the use of scientific jargon, containing terms like those cited above which produce great psychological effect upon people of science but have no precise meaning. The teaching of hypotheses and their testing certainly does not yet allow us to avoid such occurrences altogether, but it will at least make us approach such situations with greater caution and responsibility in the future.

How Science Grows

Here we must acknowledge Popper's third paradox (1970): the progress of knowledge can be presented as a process of revolutionary change and not mere accumulation. Human knowledge is not accumulated with the growth of science like books are accumulated in libraries and exhibits in museums. In the process of the development of science, the most essential elements in it, including its language, are destroyed, changed, and rebuilt. If, says Popper, in other fields of human activity we can

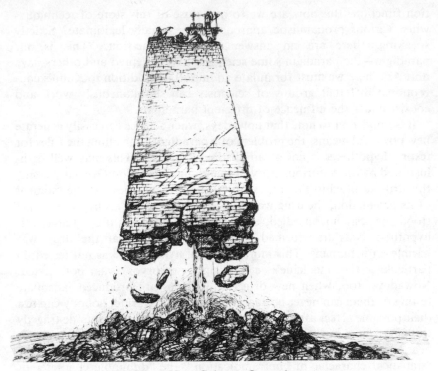

How Science Grows[5]

speak only of changes and not of progress, since a criterion of progress does not exist there, in the development of science we can speak of *real* progress. Falsifiability is the criterion of progress: old theories are replaced by new ones able to stand more severe tests. However, Popper stipulates that resistance during falsifying is not a unique merit of scientific theory; it must also possess a great explanatory and predictive power. The latter — prediction of new phenomena — is just what places a hypothesis in danger: if the predicted phenomenon is not experimentally discovered, the theory will have to be rejected. This concept of a revolutionary character of changes in scientific theories reflects what happened in the twentieth century in physics. This field of knowledge was excessively rich in theoretical conceptions which, as it seemed in the last century, were remarkable for their putative immutability, and this allowed them to be elevated to the rank of natural laws. Still within my recollection, in the large physics lecture hall of Moscow State University's old building, Newton's laws were solemnly written out as certain in-

⁵ This picture was published in the journal *Khimiya i Zhizn (Chemistry and Life)*, no. 6, 1979.

dubitable fundamentals of physics—at least they seemed so to the founder of this lecture hall, the physicist Peter Lebedev.

Popper's viewpoint may be interpreted as a concept of permanent revolution in science. It is often opposed by that of Kuhn (1970*a, b, c*) on the existence of two cycles of science development: lengthy periods of normal science and short outbreaks of scientific revolution. During the first of these cycles, science progresses quietly, proceeding from certain fundamental knowledge and methodological ideas generally accepted at the time. It is to denote the intellectual climate of this period that Kuhn introduced his theory of emphasis on the paradigm. According to him, a paradigm generates scientific collectives built as closed communities where critical analysis is forbidden.

I do not believe it reasonable to stress a strong opposition of these two outlooks. They describe the same phenomena by certain metaphors. On a large time scale, we perceive the development of science as a continuous evolution; on a small time scale, as a creation of separate closed collectives, often turning off the highway and degenerating in their isolation. But in closed collectives, too, we often observe hidden struggles, and if these are absent, then we shall speak of the danger of "normal" science, as it was put by Popper (1970).

However, another approach for describing the process of the development of science is possible. In my study of the science-of-science (Nalimov and Mul'chenko, 1969), I followed the lead of Derek Price and constructed simple cumulative curves representing growth of the number of scientific publications, estimating the citations of publications, etc. Here again we deal not with an inconsistency but with different approaches or, to be more precise, with a view of one and the same phenomenon from different aspects. The science-of-science approach allows us to comprehend and describe many interesting phenomena connected with the development of science as an information system. By the way, the idea of "invisible colleges" which emerged in science-of-science studies is just a description of how scientists of many countries are united under the aegis of one paradigm. But in such an approach, the logic of scientific development escapes the attention of researchers. The more thorough approach of Popper–Kuhn calls attention to this, but still we get to a difficult position: revolutionary changes in the development of science are very hard to express quantitatively. We no longer have an instrument with the help of which we can watch the development of an informational system. It is interesting to note that through history there have existed information systems which did not change in a revolutionary way, e.g., medieval theology; one of these which has been preserved up to now is the system of Yoga. It is inexplicable why nobody has so far studied the life of such systems as a contrast to the development of science. We cannot answer with confidence even the question of whether the exponential growth of publications, so specific for science, is a consequence of revolutionary changes in its initial premises.

Is the Probabilistic Evaluation of Hypotheses Possible?

The next step in Popper's conception is opposition to the broadly accepted tendency in science, originating with Laplace, to speak not of the correctness of a hypothesis but of its probability. If, says Popper, we regard the progress of science as the emergence of theories with increasing content, it should immediately follow that their probability decreases. To illustrate this he gives the following example. Let us denote by *a* the proposition "It will rain on Friday" and by *b* the proposition "We shall have fair weather on Sunday"; then *ab* will be the statement "On Friday it will rain, and on Sunday we shall have fair weather." It is obvious that the content of the conjunction *ab* of two propositions *a* and *b* will always be greater than or at least equal to that of the separate components, and the probability of the emergence of joint events will always be less than or equal to that of the emergence of separate events. It may be symbolically expressed as

What is Truth?

$$ct(a) \leqslant ct(ab) \geqslant ct(b)$$

$$p(a) \geqslant p(ab) \leqslant p(b)$$

where $ct(a)$ represents the content of statement a and $p(a)$ is the probability of event a. Hence, it follows that the increase in the proposition's content is accompanied by the decrease of its probability. Popper is apt to consider this simultaneously trite and fundamental result as a discovery.

This fourth paradox of Popper can be substantially criticized. I believe that here a very important idea is expressed without sufficient accuracy, whence comes the possibility of its erroneous comprehension. The point is that it makes sense to speak of the probability of an event only when the space of elementary events is given with sufficient unambiguity. If this is ignored, false paradoxes arise immediately. As an example, consider the paradox of von Mises as it is presented by Tutubalin (1972):

> In the classical probability theory there is the definition, "two events are called incompatible if they cannot occur simultaneously," and the theorem, "the probability of the sum of two incompatible events equals the sum of their probabilities." R. von Mises invented the following paradox: a tennis player can go to a tournament either in Moscow or in London, the tournaments taking place simultaneously. The probability of his winning the first prize in Moscow is 0.9 (of course, if he goes there); in London it is 0.6. What is the probability of his winning the first prize at all? Solution: according to the classical theory, the two events are incompatible, and, therefore, the probability in question is $0.9 + 0.6 = 1.5$.

This paradox is, as a matter of fact, a result of misunderstanding since the probabilities 0.9 and 0.6 relate to *different spaces* of elementary events.

Now let us return to Popper's paradox. If we deal with the probability of a serious scientific hypothesis, we have to mention the space of statements for which the probability can be estimated; otherwise, the results will be meaningless. A new revolutionary theory emerges on the intellectual field formed by a significantly different, previously existing theory. If the probability of a new theory is estimated in the space of statements given by the previous theory, its probability will obviously prove very small: the smaller the probability, the more revolutionary it looks. If we trace the process of the development of science, we shall see that the most promising and fruitful scientific hypotheses at the moment of their creation arouse frantic opposition in scientific circles. This means that they were regarded as having a small probability from the standpoint of their intellectual background. The resistance of scientists to novel scientific ideas is well described in the article by Barber (1961; see also Duncan, 1974).

If a newly born theory were a mere logical consequence of previously made statements whose truthfulness was indubitable, we could ascribe to it a unity probability; however, it is hardly pertinent at all to use probabilistic language in this case. Now assume that the new hypothesis has predicted new effects not resulting from the old one, and they have successfully been discovered. The prestige of the theory will immediately rise, and it will determine a further line of research. Around such a theory, a definite intellectual field will be formed, and its probability will increase in the space created by these new statements; later, the probability may start decreasing again. The latter turns out to be unstable since the field itself of statements where it is found proves such. It is in this sense that I think it reasonable to oppose the ascribing of probability to scientific hypotheses.

However, it is well known that one of the tasks of mathematical statistics is precisely that of estimating the probabilities of hypotheses. But here the space of elementary events is built differently. Let experiments be carried out in chemical kinetics, aimed at choosing one of the two rival models (or hypotheses). The research ends when, say, one of the models can be ascribed 0.99 probability and the other one 0.01. But high probability ascribed to one of the models relates only to the set of elementary events, connected with testing these two models. In another problem formulation when several more rival models are taken into consideration, the probability of the winning model may happen to be fairly small.

Thus, if we understand the meaning of the concept of "hypothesis probability" as a definite and narrow one, it may be useful. And it is only natural that when a new and unusual hypothesis emerges its probability estimated on the space of all previous statements will be small. This is equivalent to stating that the hypothesis is unpredicted and revolutionary.

Our interpretation might be regarded as a semantic reformulation of Popper's fourth paradox. Indeed, the more unexpected a new hypothesis is, the more meaningful it is, and the less its probability is at the moment of its appearance; the probability of the withering hypothesis at its last stage of existence is high, but the value ascribed to it is small since it has by now lost its heuristic power.

Finally, I would like to draw the reader's attention to a consequence of the statement that the probability of a new powerful hypothesis at the moment of its emergence *must* be small. What I have in mind is the notorious problem of forecasting the development of science or of scientific–technological progress. It is often said that scientific forecasting of the development of science is as natural as predicting, by means of scientific hypotheses, new phenomena in the world of physical experiments. As a matter of fact, the analogy is false, since forecasting cannot base itself upon the experiment in decision making. At the moment of generating forecasts, the forecaster has to choose among a set of hypotheses with low probability, not having at his disposal the results of a crucial experiment. The latter are obtained only when the forecasts come true, i.e., when they are no longer forecasts.

What Is Science

Any classification is conventional and does not, certainly, reflect anything but our personal viewpoint, the position from which we regard a complex system. Every such viewpoint gives rise to a specific concept or a metaphor which behaves in some sense like the system described and in another sense otherwise. But without classifying we cannot think.

Proceeding from the concept of falsifiability, Popper easily draws a demarcation line between science and nonscientific, metaphysical statements. Only those statements prove to be scientific hypotheses which can be placed in danger of rejection through testing. This demarcation criterion can by no means be regarded as unconditional. There may exist theories that are well suited to testing, as well as weakly testable and untestable ones, the last being of no interest for empirical sciences. In any case, all philosophical constructions without exception, according to Popper, fall in the class of non-scientific ones, though he acknowledges the possibility of their meaningful discussion. Freud's concepts, too, belong in the same class.

From the standpoint of a scientist, the demarcation line suggested by Popper at first sight looks rather natural. Indeed, a scientist cannot be interested in a hypothesis if it is formulated so as not to be testable with the risk of its rejection. Hence comes the distrust of all experimentally untestable statements. On the other hand, if we accept unconditionally Popper's point of view, then not only astrology but his own conception will get into metaphysics as well. Mathematics seems to get there, too. Finally, some concepts of natural science do not yield to immediate falsification, e.g., theoretical constructions in astrophysics or the theory of the origin of the genetic code. I believe it more reasonable to manifest a certain reticence and to speak, on the one hand, of science *in the narrow sense* of the word, science separated from other types of human intellectual activity by Popper's barrier, and of science *in the broad sense* of the word, including here purely logical constructions — mathematics and formal logic, with their specific concepts of the correctness of judgments, and constructions like astrophysical or biologico–evolutionary hypotheses. They all are based, to some extent, upon statements yielding to falsification. Finally, the concepts of a cosmic character cannot be put under the conditions of falsification, and for this reason they do not relate to science in the narrow sense, but in the broad sense they must be considered as scientific structures since, as opposed to myths, they emerge in a definite scientific–intellectual field which is created by science in the narrow sense. Then Popper's concept does not fall into the same class as astrology and chiromancy. True, in this approach, Popper's constructions (probably he will not like this at all) will be juxtaposed to existentialism, since despite all its iconoclasm, existentialism originated in the intellecutal field created by modern science.

The sense of any classification is justified only by the convenience of its application for constructing a system of assertions. If we are to accept the scheme outlined above, we shall clearly see that Freud's theories, though related to science in the broad sense, have not the same status as, for example, the special and general theory of relativity. There often emerges the question: is a scientific forecast of scientific–technological progress or, even wider, of the development of our society and culture possible? The answer to this must be negative: scientific (in the narrow sense) forecasting of social phenomena is impossible — we cannot at present subject to verification or, better, to falsification what will happen in the distant future. And what cannot be tested by a crucial experiment cannot be considered scientific. However, being situated in a certain intellectual field created by science in the narrow sense, we can speak of the future in terms which will be of greater meaning for us than astrological predictions, at least because it is more fruitful to discuss them.

The key to separating scientific and non-scientific conceptions should rather be their capacity for self-development, i.e., for self-destruction. If you like, this is a dialectical definition of science. However, this is only a necessary condition, not a sufficient one: we can point to religious systems which, in the process of evolution, changed beyond recognition. How can necessary and sufficient conditions be formulated? There is no answer to this question.

How Can the Epistemological Role in Discoveries Be Estimated?

Scientific hypotheses, at least some of them, may predict essentially new and previously unknown effects. It is in the possibility that such predictions will stimulate precisely directed activity of the researcher that the principal power of scientific theories, and sometimes the criterion of their correctness, is found by many. Indeed, we know that a lot of effects in science, especially in physics, were discovered not by chance, but as a result of theoretically directed research. However, the following question still remains open: What epistemological value can be ascribed to discoveries in natural sciences? If hypotheses are only guesses sequentially replacing one another and not true cognition of nature in some indubitable and strict sense, then discoveries made by means of these guesses are probably to be interpreted also not as links in the progress of cognition but as consistent and more and more profound mastery of nature. The history of culture supplies us with many examples of significant mastery of nature achieved on the basis of fairly odd, from the modern standpoint, theoretical constructions. The first example is the very profound mastery of the human body in Yoga, though its theoretical physiological ideas now seem very primitive. Is it possible to state that their brilliant achievements, which can be interpreted as supported by practice, are enough to justify recognition of their general philosophical structures as possessing a great epistemological power? Yoga's concepts answered all the requirements mandated for scientific hypotheses: they had heuristic power, giving impetus to the discovery of new physiological aspects of the human body; they were formulated so as to make possible their testing under rigid conditions for falsification; and they withstood the test. Nevertheless, modern science cannot acknowledge that Yoga is a step toward the truth. Another example is the culture of Egypt: there, amazing technical achievements were stimulated by altogether odd ideological structures. We are far from calling them theories in the modern meaning of the word, but, again, they were at least certain guesses about the World and its arrangement, which in some queer way served to stimulate technological progress.

Therefore, is not it wiser to be cautious and ascribe to what we call scientific discoveries no more than a status of mastering nature? Theories can stimulate this process in a greater or lesser degree, which may be a measure of the heuristic power of the theory. But why should it simultaneously be a measure of its epistemological power? Having given up the religious approach, are we not ascribing to human beings what has previously been naturally ascribed to gods, the creators of worlds? If we believe in the evolutionary development of the intellect, it seems natural to believe that this process originated from the desire to master nature rather than to cognize it. Perhaps what we mean when we speak of cognition is in fact the mastery of nature (since we are ignorant of the true meaning to be put into the concept of cognition), and this will put an end to all arguments. But scholars are confident that science has epistemological power. This confidence is a constituent of our modern paradigm. The question "Why are they confident?" is forbidden in the frame of this paradigm. The philosophy of science attempted to violate this code of manners by formulating the question, but it could not give a convincing answer.

For those who study the philosophy of science, it is extremely interesting to trace the development of alchemy. This curious practice existed for about 1,000 or 1,500 years. It was a gigantic, technically well-equipped experiment aimed at mastering nature. Its results were presented in a cookbook form and comprehended in poetico-mythological formulations. There was no theory in the modern sense of the word since there did not exist the language of abstract symbols by means of which it would be possible to formulate results in a general and, consequently, theoretical form. Still, obvious progress was made, though slowly: hydrochloric and nitric acids, a number of elements, and certain reactions were discovered; gunpowder was invented. These are real discoveries of scientific significance. Alchemy, as an epoch of European pre-scientific life, has not so far been completely understood. What stimulated its discoveries? Were they started by mythological structures uncomprehended by us? A vivid description of alchemy is given in the article by Rabinovich (1973). All of this leads me to ask: Do not some kinds of our scientific activity which have their roots in the distant past (e.g., medicine with its cookbook decisions) seem to resemble the alchemical approach?

Some Historical Parallels and the Principal Consequence of Popper's Concept of the Growth of Knowledge

If we want to uncover the contradictions in a philosophical tradition, it must be analyzed against a background of ideas that are similar but generated in a quite different intellectual field. Popper's concept is a concluding link in the long chain of European rationalism beginning with the Hellenic world. Therefore, it seems natural to challenge it not through European but through Oriental traditions of rationalistic criticism.

One of the early schools of logical criticism in ancient India was the philosophy of Jainism, whose emergence dates back to the sixth century B.C. [For greater detail on Jainism, see the paper by Mahalanobis (1954), who tries to demonstrate the closeness of these ideas to modern probabilistic notions.] Contemporary authors describing this system define it as the philosophy of non-absolutism, pluralism, relativism. In this philosophical system, on the basis of logical analysis of nature, it is stated that the correctness of any proposition is only conventional: an opposite and contradictory proposition can always be recognized as true in another sense and proceeding from different grounds; reality can be regarded from various angles. Following Hutten (1956), we can express this idea in modern language like this: theories in science are only metaphors; they generate models which behave similarly to but not exactly the same as the phenomena described by them.

Logical nihilism was expressed even more violently by Nāgārjuna (in the beginning of the twentieth century), the representative of the philosophy of Mādhyamika (middle way), which derives from the teaching of Buddha. [An idea of Nāgārjuna's logic of judgments can be obtained by reading the English version of his tractatus (Bhattacharya, 1971)]. By a sequential chain of precise logical judgments, Nāgārjuna comes to the following conclusions: a thought cannot cognize either itself or something else; truth is unexpressible; knowledge is impossible; there is no difference between truth and delusion; the world of experience is illusory. By means of logical analysis, Nāgārjuna tries to prove the impossibility of building empirical knowledge. His results are summarized by Radhakrishnan (1962) in his well-known monograph *Indian Philosophy*:

> Nāgārjuna exhibits the conditions which render experience possible, shows their unintelligibility, and infers the non-ultimate character of experience. (p. 656)

> The world of experience is bound by the relations of subject and object, substance and attribute, actor and action, existence and non-existence, origination, duration and destruction, unity and plurality, whole and part, bondage and release, relations of time, relations of space; and Nāgārjuna examines every one of these relations and exposes their contradictions. If non-contradiction is the test of reality, then the world of experience is not real. (pp. 697–698)

However, Nāgārjuna is still on the middle way. He does not reject the truths of the mind even if they are not ultimate. Knowledge of practical truth proves for him the way to transcendental knowledge.

The cautious relation to knowledge in ancient India is well illustrated by the following quotation from Isha Upanishad (Va'yasane'ya Sanhita, 1957):

> 9. Those who worship ignorance, enter into gloomy darkness, into
>   still greater darkness those who are devoted to knowledge.

True, we should remember that the concept of knowledge in India at that epoch was different from the modern Western one, since it was related to the ethico–applied trend of thought.

Another element remains to be mentioned: the ever-repeated refrain of ancient Indian philosophy that knowledge is only destruction of ignorance. The liberation from ignorance is the way to nirvana.

Now let us see how Popper's critical realism will look against the background of the rational nihilism of ancient India. Both trends become aware of logical difficulties relating to posterior synthetic judgments, but these difficulties are solved differently. Nāgārjuna shares complete nihilism, the confidence in the impossibility of constructing empirical knowledge. Popper's view is not nihilism but criticism, the understanding that scientific concepts are only conjectures; in no way are they structures deduced from experience. Experience has another role, that of falsifying hypotheses. In this case all troubles coming from the logic of constructing knowledge from experience are removed. True, the most important question — the creative process — remains unexplained by Popper: it is merely relegated to the class of conjectures, i.e., alogical procedures. As a matter of fact, Popper's entire concept is only a factographical description of what is going on in European science. Here the following idea, though a bit paradoxical, might be formulated: Indian thinkers proved too consistent and could not allow the inconsistent approach to the development of science taken by European thought.

But most interesting is a juxtaposition with the idea of knowledge as destruction of ignorance. Popper's concept can be expressed in similar terms. If the growth of science is not a mere cumulation of knowledge but permanent creation of a new hypothesis rejecting the previous ones, then it is no other than a consistent process of destroying previously existing ignorance. It is here that the most significant difference from the ancient Indian concept lies: Nāgārjuna limited the fight against ignorance to the development of criticism; he did not try to construct any positive concept since he saw only too well that in doing so he would have to face the same weak points he himself criticized so effectively. In Popper's concept, the fight against ignorance is a chain of constructing stronger and stronger ignorance. The justification of this roundabout progress is the process of mastering nature that accompanies it. But the process of cognition turns into a series which does not necessarily converge to true knowledge, even if the process of cognition could be thought of as continuing to infinity. Worse than that, I believe that the process of cognition cannot go on endlessly without stopping. At every step old ignorance is destroyed by constructing new knowledge, but from the viewpoint of the future it turns out to be even stronger ignorance which it becomes harder and harder to destroy, as time goes by. Is this not precisely the state which many physicists have now achieved, especially in the theory of elementary particles? Old concepts in physics prove insufficient both for a profound comprehension of intensively accumulating new experimental data and for the prediction of novel effects. At the same time these concepts are powerful enough to oppose their revolutionary change. Here is how the well-known physicist L. A.

Artsimovich (1969) picturesquely describes the situation in a popular-scientific article:

> They have so far been saved from the most dangerous disease called "Crisis of Genre" (it consists in the disappearance of fruitful scientific problems; in research institutes caught by the disease, scientific workers are bursting with energy while the directors pass sleepless nights pondering where to direct the unused energies of a large collective) only by technological applications perpetually increasing in variety and practical value.

The prolonged crisis in theoretical physics is an acknowledged fact. Certainly, the construction of such all-embracing and, therefore, unavoidably cumbersome theories that occurs in physics does not take place in all fields of knowledge. In biology, psychology, sociology, linguistics, and partially even in chemistry, mathematics, if used, serves to construct only local models and not such all-embracing conceptions as quantum mechanics or the theory of relativity. Local models are but a view of complex phenomena from one of all possible angles. We have already mentioned that such models have the status of a metaphor: in some sense they behave like the phenomenon described but in another sense they behave differently. The construction of a new model does not necessarily result in rejection of the previously developed one (see Kleene, 1952). Several models may exist simultaneously, forming a mosaic structure which reflects various aspects of the phenomenon under study. How should we in this case regard the development of science on a large time scale; as an infinite growth of a mosaic whose elements do not develop inner coherence? Hardly so. It seems likely that here, too, we can expect a revolutionary change which would then relate not to the rejection of any particular model but to giving up such an approach completely and replacing it with some other as yet unknown. Here the same difficulty will arise: the better the method of constructing local models becomes, and the more results they yield, the more difficult it will be to transcend them.

But let us return to Popper's conception of scientific growth. It ought to result in a certain refined agnosticism, though Popper himself does not come to this conclusion. Popper does not even think himself a relativist, as can be seen in the following quotations (Popper, 1970).

> I am not a relativist: I do believe in absolute or objective truth, in Tarski's sense (although I am, of course, not an absolutist in the sense of thinking that I, or anybody else, has the truth in his pocket) . . .
>
> I do admit that at any moment we are prisoners caught in the framework of our theories; our expectations; our past experience; our language. But we are prisoners in a Pickwickian sense; if we try

Science as a Game[6]

---

[6] This picture was published in the journal *Khimiya i Zhizn (Chemistry and Life)*, no. 1, 1978.

we can break out of our framework at any time. Admittedly, we shall find ourselves again in a framework, but it will be a better and roomier one; and we can at any moment break out of it again.

I fail to comprehend Popper's optimism, though logically it follows from his premises. I am tempted to ask the question: Did not some cultures (say, Egyptian) perish and certain outstanding trends of thinking (say, ancient Indian) decay as a consequence of their achieving the level of ignorance (in the latter case expressed, for example, in extreme logical nihilism) which did not allow the possibility of destroying it? Who knows to what extent the power of ignorance in European knowledge will be conservative?

A Metaobserver's Glance at Science

Imagine that the Earth were visited by a metaobserver from another world, free from the prejudices of our conceptions of science. I think his report would probably look like this:

> Science is a kind of Game. The Game has special rules which are known and clear to everybody though they have never been classified and codified. The rules have been basically constant for about 300 years. In the process of the Game, ingenious and ever more complicated theoretical structures are created, but the players do not seem to believe them absolutely. Anyway, they perceive ultimate knowledge as a delusion since only the scholar who manages to destroy what has previously been created is considered truly gifted. What is the prize in the Game? This is not quite obvious. For some, it is the ability to build a more ingenious theory; for others, and for those not directly participating in the game, it is the ability to master previously unknown powers of nature, which they unaccountably succeed in doing by proceeding from their ephemeral theories; for still others, it is the ability to get hold of something purely material. The last play the same Game, but they play according to quite different rules and interfere with other people. One of the principal rules seems to stipulate that the Game must not be dull. The moment it loses its acuteness, more ingenious conjectures start arising, the rules are modified, and, surprisingly, again everything is all right, though it becomes more and more difficult to play.

Throughout this chapter I have made a lot of remarks in connection with the statements of Sir Karl Popper. Now I would like to make one more: to express my profound admiration of the amazing precision of his structures.

Chapter 2

Scientific Creativity as a Manifestation of Intellectual Rebellion[1]

A Bayesian Approach to the Problem

Our knowledge of thinking, especially of creative thinking, is limited. But if this knowledge is formulated as a model, it will help us to discuss meaningfully the ways of forming conditions favorable for the development of creative scientific activity. Below I make an attempt to use the Baycsian theorem to construct an outline of a model of thinking.

However meagre our knowledge of thinking may be, we are still able to state one thing with certainty: when in our everyday reasoning we try to ascribe correctness to a proposition we rely not so much on the rules of formal logic as on the correspondence of the new proposition to our previous knowledge and past experience. The new proposition is acknowledged to be true if it strengthens what we have already vaguely guessed. It seems natural to speak at present not so much of the logic as of the psychology of thinking since the correctness of a proposition is stated at a psychological level. Here I to some degree fall in line with the statements of those existentialists who derive from Kierkegaard. Later, I shall try to give a new interpretation to the mechanism of the personal perception of correctness.

The degree of correspondence of a new judgment with previous prerequisites can naturally be of only a probabilistic nature. And if this is so, then in order to build a model giving the correctness of a judgment, we can make use of the Bayesian theorem.[2] Our existing concept of some

---

[1] This chapter was published in Russian in the journal *Izobretatel' i Ratsionalizator*, no. 7, 1976. It was translated by A. V. Yarkho.

[2] I earlier used the Bayesian theorem to build a model giving the mechanism of comprehending phrases

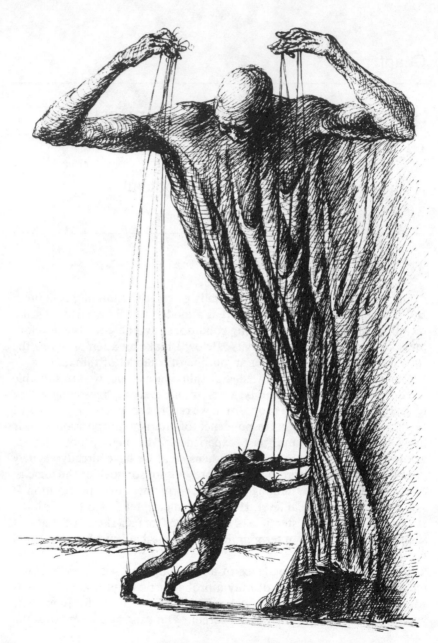

Intellectual Rebellion

problem can be presented probabilistically as a prior distribution function $p(\mu)$, which means that various judgments of the problem arranged in some way on the scale μ are ascribed different probabilities. In speaking of scientific activity, this is to say that the prior distribution function $p(\mu)$ gives the paradigm accepted at some moment in time. Further, we can speak of the probabilistic presentation of judgments of the phenomenon related to the problem μ. This conditional probability is put down as $p(y|\mu)$. Then the ultimate posterior judgment $p(\mu|y)$ obtained when the new proposition is weighed according to the system of old concepts can be presented as it follows from the Bayesian theorem:

$$p(\mu|y) = kp(\mu)p(y|\mu)$$

where k is merely a constant of normalization.

Note that the posterior distribution function obtained in the nth reasoning becomes prior with respect to the next $(n + 1)$th one. The conceptions are prior in a statistical sense, not in the Kantian sense.

Now I would like to illuminate by examples the way in which this model can be used to explain how correctness can be ascribed with great probability to statements not justified from the viewpoint of a formal logical analysis.

Consider two statements (Box, 1974):

1. Statistical data show that the death rate caused by lung cancer of smokers is higher than of non-smokers. So it seems natural to conclude that if a group of people stops smoking, then the death rate from lung cancer within it will decrease.

2. Statistical data show an almost perfect straight line relationship between the increasing number of storks' nests observed each year in some German towns and human birth rate for that year. But it will occur to nobody to conclude that the over-population problem can be solved by destroying storks' nests.

From comparing the second judgment with the first one, it seems to follow that the statement about the harmful consequences of smoking is as unjustified as that about the demographic role of storks' nests; in all probability, this is also the opinion of the author of the paper from which I have borrowed the comparison. Indeed, in the second case we deal only with a correlation and in no way with a causal relation: the increases of both the stork population and the human population are due to one and the same hidden (for this research) cause, that of well-being. In the first case, strictly speaking, only the correlation is statistically

---

constructed over words with fuzzy semantics (Nalimov, 1974b, 1981). I also used the same theorem to construct a model of human social behavior (Nalimov, 1973; see also pp. 225–233 of this book).

valid; it may well be that there exists a hidden cause, perhaps of genetic
origin, responsible for both the need to smoke and the susceptibility to
cancer.

From the formally logical standpoint both the above-cited statistical
observations are obviously insufficient to allow any serious recommendations. However, on the psychological level we are prone to ascribe a
high degree of legitimacy to the recommendation: stop smoking in order
to decrease the death rate from lung cancer. The above-mentioned model
based on the Bayesian theorem reveals the mechanism of such a judgment. Indeed, a priori we know much about the relation between cancer
and the effect of resins, tars, etc., and we are apt to ascribe to these relations a causal character rather than a correlational one. But if now we
consider thoroughly the statistically discovered relation between smoking
and increased death rate from lung cancer, we must ascribe all this with a
high probability to purely correlational links (strictly speaking, interpretation of statistical relations cannot claim anything more), but with a
low probability we can suppose a causal relation behind the statistical
observations. In Bayesian terms we deal here with two distribution func-
tions: $p(\mu)$ and $p(y|\mu)$, where μ denotes the cancer problem in its broad
interpretation and y stands for the problem of the possible effect of
smoking upon the progress of cancer; function $p(y|\mu)$ will reflect the
probabilistic character of our judgments about statistical data related to
the smoking–cancer problem before we compare them with prior
theoretical concepts. After multiplying these two functions and normalizing the square below them to unity, we shall obtain, according to
the Bayesian theorem, the posterior distribution function $p(\mu|y)$ showing
how our probabilistic estimations of lung cancer causes changed when we
began to consider them in connection with smoking. These considera-
tions make it clear that function $p(\mu|y)$ may turn out such that we shall
ascribe a sufficiently large probability to the statement of a causal relation between smoking and lung cancer. Nothing of the kind will happen
in the case of storks' nests and birth rate. All our prior judgments on
birth rate are obviously such that we are apt to ascribe zero probability to
the causal dependence of birth rate on the number of bird nests.

Now let us return to the problem of scientific creativity. In Bayesian
terms, the first and foremost thing in a truly creative process is construct-
ing the function $p(y|\mu)$ connected with formulating an essentially new
and original idea in discussing the problem μ. We are absolutely ignorant
of the mechanism of this phenomenon. Following Karl Popper (1965),
we can speak of this stage of creativity only as an insight whose appearance in the consciousness of a researcher does not differ from the
process of myth creation. Later comes another stage of the creative process: acknowledgment of the new idea on the background of the para-

digm existing at a certain moment in the scientific community. The mechanism of this stage may well be described by the Bayesian model of thinking. A genuinely scientific discovery will be an appearance of a new idea y that has led to the formation of a new paradigm $p(\mu_2) = p(\mu_1|y)$ which could not have emerged on the background of the previously acknowledged paradigm $p(\mu_0)$. The scientist who has made a discovery

Scientific Creativity

should not only have an insight connected with the new idea y, but he should also have a special, far less rigid prior distribution function $p(\mu_1)$ different from the existing paradigm $p(\mu_0)$.

This can be illustrated by Einstein's discovery of special relativity theory. When all is said and done, this theory is nothing more than quite an unusual, though logically precise, comprehension of Lorentz's transformations. Why could not Lorentz, an outstanding physicist of his time, do it himself? Why had not Poincaré succeeded, or Hadamard, another gifted mathematician who also tried to understand the physical sense of Lorentz's transformations? Here is what Hadamard (1949) himself writes on the point:

> Absolute differential calculus is closely connected with the theory of relativity; and in this connection, I must confess that, having observed that the equation of propagation of light is invariant under a set of transformations (what is now known as Lorentz's group) by which space and time are combined together, I added that "such transformations are obviously devoid of physical meaning." Now, these transformations, supposedly without any physical meaning, are the base of Einstein's theory. (p. 52)

From these lines it clearly follows that the special relativity theory was created by Einstein and not by any of the scientists interested in the physical sense of Lorentz's transformations because his prior distribution function allowed wider possibilities than the existing paradigm. In this sense Einstein's interpretation of Lorentz's equations was deeply subjective. It became objective only when, as a consequence of the facts observed, it acquired general popularity and gave rise to a new paradigm. The appearance of relativity theory was an intellectual rebellion against the existing paradigm of "classical physics."

Another example again refers to relativity theory. In the recently published Russian translation of *Relativity, Thermodynamics, and Cosmology* by R. Tolman (1934), well known to physicists, one finds the following words concerning the formula $E = mc^2$:

> This relation, which would imply an enormous store of energy m_0c^2 still resident in a particle even when it is brought to rest, appears somewhat more strained than our previous considerations, but nevertheless logically plausible. (p. 49)

In a note the editor of the Russian edition remarks: "This was written in the years when nuclear physics was in the bud." We see that even many years after the appearance of the relativity theory the pressure of the "classical physics" paradigm was still so strong that one had to make stipulations as to what seemed logically correct within the system of concepts of relativity physics. The paradigm changed radically only after the first atomic bomb explosion. The conceptual secret was revealed. And since there was no longer psychological inhibition, the only thing to do was to "overtake" and "outstrip." The situation was analogous after the launching of the first manned spaceship.

An excellent example of the paradigmatic pressure is the well known collection of papers *Philosophical Problems of Modern Physics* (Maksimov, 1952). The article by I. V. Kuznetsov states:

> One of the most significant tasks of the Soviet physicists and philosophers is the revelation of Einstein's reactionary ideas in physical science. (p. 47)

In the article by A. A. Maksimov from the same collection, one can read:

> The equally absurd anti-scientific speculations are the theory of an expanding universe having a beginning and end in time, an idealistic interpretation of space, time, and simultaneity preached by Einstein, the principle of complementarity by Bohr–Heisenberg, as well as the conception of resonant structures in the so-called theory of resonance which does not correspond to anything objectively existing in nature. (p. 177)

All this may be very interesting and even instructive for those who study the history of science and the psychology of creative activity.

We can certainly speak of creativity of another kind: of sub-discoveries made within the frame of existing paradigms. I believe, for instance, that the discovery of the Raman effect, despite all its immense significance for physics, was made within the existing paradigm formed by quantum mechanics.

It seems pertinent now to ask a question: Whom do we graduate from our higher educational institutions if we present only rigid paradigms of science — students capable of creative activity of the first kind or only of the second one?

If the above-exposed considerations are recognized as worth a serious discussion, it would be useful to consider the problem of teaching: In what way should we teach in order not to form too rigid paradigms in the minds of students or, moreover, in order that students will not be afraid to pass the limits of the paradigms. At our present level of ignorance concerning the mechanisms of the creative process, this is the basic problem of education.

Chapter 3

Mathematics as a Language of Science[1]

Using Mathematics to Describe the External World

> *The two elements can be distinguished by reflective thought, but cannot be rent asunder.*
>
> SATKARI MOKERJEE[2]

Introduction: Formulation of the Problem

The language of mathematics is an oft-repeated word combination, a sort of language cliché.[3] I would like to analyze this phrase and reveal its content.

It is convenient to start studying a phenomenon where it acquires its most explicit form. Mathematics as a language gets especially explicit while describing the phenomena of the external world. To my mind, *applied mathematics is use of mathematics as a language*. It is also possible to speak of the *language* of mathematics while discussing mathematical problems proper. But this delicate question is beyond our consideration. I should like only to demonstrate that mathematics, as a deductive science with its own problems, turns into a language when it starts to be used for describing external phenomena.

To succeed in this task, we must separate pure and applied

---

[1] The contents of this chapter in a slightly modified form entitled "Logical Foundations of Applied Mathematics" were published in Russian as a preprint of Moscow State University (no. 24, 1971) and in English in the journal *Synthèse* (27:211–250, 1974). This chapter was translated by A. V. Yarkho.

[2] From *The Jaina Philosophy of Non-Absolutism* (Calcutta, 1944), cited by Mahalanobis (1954).

[3] The linguistic nature of mathematics was well comprehended by Kant in whose system of views mathematical judgments were prior synthetic judgments preceding experience. Further, he held that, if one confined himself to "pure" rational concepts of natural science, each separate teaching of the nature of science would be as rich as the mathematics it contained.

mathematics. First of all, I should like to show that the internal contents of mathematics, *its structures,* turn merely into grammar when used to solve applied problems. Is there any sense in drawing a demarcation line between applied and pure mathematics? The question can be formulated more broadly: What is the sense of any system of classification? Is there any sense, for instance, in dividing people into men and women and in speaking of the peculiarities of male and female psychology, or is it possible to speak only of human psychology and its aspects? Classification of the phenomena observed is a mode of human thinking. It is not at all a simple task, and it may not always be worthwhile to discuss the true nature of the elements of classification. In many cases the accepted system of classification is only an idiosyncratic point of view which we ourselves choose in order to consider a complex system. A look at the system from a certain point of view may give rise to biased notions. This was wittily pointed out by S. I. Vavilov (1955) in his book *The Eye and the Sun*: on page 93 he showed a picture of a reclining man, taken with the camera near his boots. The boots showed up immensely large and the man's head excessively small. The picture of the man proved to be distorted though it had been received by means of an optical objective, which by virtue of its name should have given an objective picture. It is often very tempting to look at a system from a specific angle emphasizing those sides of phenomena which usually remain hidden. This is the aim of my attempt to draw a demarcation line between pure and applied mathematics. Many mathematicians furiously oppose such a division, saying that no special applied mathematics exists but only applications of mathematics. Such a formulation is due to a desire to avoid discussing the real problems which gave birth to the difficulties we face when we attempt to use mathematics broadly in all the variety of scientific and industrial research. A curtain of diffidence is thrown over all possible aspects of a complicated and, probably, indivisible phenomenon.

In the following two sections I shall try to give a brief account of the modern ideas of the logical foundations of pure mathematics. We need this to create a background for our further judgments on the logical structure of applied mathematics. The presentations will be mainly illustrated by examples from mathematical statistics, the branch of science with which I am most familiar.

Axiomatic–Deductive Construction of Traditional Mathematics: Logical Structures of Pure Mathematics

A desire to draw a demarcation line between pure and applied mathematics is evidently as old as mathematics itself. I shall restrict myself to citing Felix Klein (1945), a well-known mathematician:

> With the *Ancient Greeks* we find *a sharp separation between pure and applied mathematics,* which goes back to Plato and Aristotle. Above all, the well-known *Euclidean structure of geometry* belongs to pure mathematics. In the applied field they developed, especially, *numerical calculations,* the so-called *logistics* (λογος, general number). To be sure, the logistics was not highly regarded, and we know that this prejudice has, to a considerable extent, maintained itself to this day — mainly, it is true, in those persons only who cannot themselves calculate numerically. The low esteem for logistics may have been due in particular to its having been developed in connection with *trigonometry* and the needs of *practical surveying,* which to some did not seem sufficiently aristocratic. In spite of this fact, it may have been raised somewhat in general esteem by its applications in astronomy, which, although related to geodesy, always has been considered one of the most aristocratic fields. You see, even from these few remarks, that the Greek cultivation of science, had its sharp separation of different fields each of which was represented with its rigid logical articulation . . . (p. 80)

Later on, during more than two thousand years, the two branches of mathematics, pure and applied, sometimes merged in the papers of many-sided scientists and sometimes diverged again. I shall not trace this process, for such a historical analysis is not a part of my task, though certainly it is of great independent interest. It became possible to approach this problem from a new standpoint when an article appeared under the pseudonym Nicholas Bourbaki bearing the original title "L' Architecture des Mathématiques." In this section an attempt is made to cast a glance at the entire variety of mathematical subjects, whereas Bourbaki of course deals only with pure mathematics, but from a certain unified standpoint. The question is formulated in the following way: Is mathematics in all its modern versatility still one science? In the following quotation from "The Architecture of Mathematics," the author speaks of pure mathematicians:

> To give a general idea of the science of mathematics at the present time seems at first glance to be an almost insurmountable task because of the breadth and variety of the subject. As in all the other sciences, there has been a considerable increase in the number of mathematicians and works devoted to mathematics since the end of the 19th century. The articles devoted to pure mathematics published throughout the world in the course of a normal year cover several thousands of pages. Doubtless, not all of them are of equal value; but after pouring off the inevitable waste, it still remains true that each year mathematics is enriched by a throng of new results, is con-

stantly diversified and ramified in theories which are ceaselessly modified and remolded, confronting and combining with one another. No mathematician, even if he devoted all his time to it, would be capable today of following this development in all its detail. A number of them shut themselves up in one corner of mathematics, which they never seek to leave, and are not only almost completely ignorant of everything outside their subject, but would even be incapable of understanding the language and terminology employed by their colleagues in a specialty separated from their own. There is scarcely anyone, even among the most cultivated, who does not feel lost in certain regions of the immense mathematical world. As for those such as Poincaré or Hilbert who imprint the stamp of their genius in almost every domain, they constitute a very rare exception even among the greatest.

Thus there can be no question of giving to laymen a precise image of something that even mathematicians cannot conceive in its entirety. But it is possible to ask whether this luxuriant proliferation is the growth of a vigorously developing organism which gains more cohesion and unity from its daily growth, or whether on the contrary it is nothing but the external sign of a tendency toward more and more rapid crumbling due to the very nature of mathematics, and whether mathematics is not in the process of becoming a Tower of Babel of autonomous disciplines, isolated from one another both in their goals and in their methods, and even in their language. In a word, is the mathematics of today singular or plural? (Bourbaki, 1950)

To the question of whether mathematics is one science, very precisely formulated, a positive answer is given. It is stated that mathematics (I repeat that only pure mathematics is meant) is a unified science. Its unity is preconditioned by the system of its logical structure. A characteristic peculiarity of mathematics is the precise axiomatic–deductive method of constructing systems of judgments. Any mathematical paper is first of all characterized by the long chain of logical conclusions it contains. But, Bourbaki remarks, stringing syllogisms together is only a transforming mechanism. It can be applied to any system of premises. This is but an outward characteristic of a system or, if you like, of its language, and it does not yet reveal the system of logical structure given by postulates.

The system of postulates in mathematics is by no means a multicolored mosaic of separate initial statements. The peculiarity of mathematics lies in the fact that the systems of postulates form specific concepts, mathematical structures rich in those logical conclusions which can be deduced on their basis. Mathematical structures are applicable to a variety of elements whose nature remains unknown (Bourbaki, 1960). In order to give a structure, it is sufficient to define certain relations between these elements on the basis of a system of axioms. Mathematics thus turns out to be of the nature of a calculus. Systems of judgments are constructed there without any appeal to implicit assumptions, common sense, or free associations. The problem consists in testing that the results obtained are indeed consequences of the initial assumptions. The formulation of the problem of testing the truth of the initial axioms is senseless. But the system of axioms determining a mathematical structure should be constructed so as to be rich in logical consequences.

The ideas of universal symbolism and logical calculus can be traced back to Leibniz, though the modern precise definition of mathematics as a strictly formalized calculus became possible only after the works of Frege, Russell, and Hilbert had appeared. Kleene (1952) offers the following characterization of Hilbert's philosophical stance:

> . . . those symbols, etc. are themselves the ultimate objects, and are not being used to refer to something other than themselves. The metamathematician looks at them, not through or beyond them; thus they are objects without interpretation or meaning. (p. 64)

We may speak of algebraic structures, such as groups, bodies, and rings (Bourbaki, 1958) or, for instance, of topological structures, where the notions of neighborhood, limit, and continuity which had appeared earlier at an intuitive level, are formulated mathematically. Bourbaki (1958) defines algebra as follows:

> The object of Algebra is the study of structures determined by there being given one or several laws of composition, internal or external, between the elements of one of several ensembles. (p. 41)

I shall dwell on topological structures so that the reader can have a clearer idea of the way mathematical structures are created. Topology (the science of location), founded by Riemann, resulted from the desire to develop a study of continuous values not on the basis of measuring distances, since in this case it is always necessary to introduce a definition of measure and to trouble about scales, but on the basis of relations of mutual arrangement and inclusion. It turned out that the notion of a "neighborhood," which is of great importance in mathematics, can be defined without resorting to the notion of distance. In order to do this, it sufficed to formulate a seemingly quite simple statement that every subset containing a neighborhood of a point A is a neighborhood of this point as well, and that the intersection of neighborhoods of the point A is also a neighborhood of this point. I shall cite this statement in the refined form it has in *Topologie Générale*, one volume of *Les Structures Fondamentales de l'Analyse* (Bourbaki, 1958):

> Definition 1 — An ensemble $\mathfrak{J}$ of parts of an ensemble E defines on E a topological structure (or more briefly a topology) if it possesses the following properties (called the axioms of topological structure): (O_1) Each union of ensembles of $\mathfrak{J}$ is an ensemble of $\mathfrak{J}$; (O_{11}) each

intersection of ensembles of ℑ is an ensemble of ℑ. The ensembles of ℑ are called open ensembles of the topological structure defined on ℑ.

Definition 2 — An ensemble possessing a topological structure is called a topological space.

Definition 3 — In a topological space E, one calls a neighborhood A of E each ensemble which contains an open ensemble containing A. The neighborhoods of a part $\{X\}$ reduced to a single point are also called neighborhoods of point X.

Thus, in a fairly simple way, the axioms of topological structures and definitions connected with them can be formulated. The choice of these axioms is, of course, arbitrary to a certain extent and appeared historically as a result of a long search. These, and likewise certain other axioms and definitions formulated in an equally simple manner, gave rise to numerous consequences inferred from them by a purely deductive method.

The formalization of mathematics, its axiomatic construction, was completed to a considerable extent by the end of the nineteenth century. During that period the German mathematician Dedekind and the Italian mathematician Peano formulated an explicit system of axioms for arithmetic, and the German mathematician Georg Cantor constructed the theory of infinite sets which appeared to be the basis for the foundations of calculus. But is was only in the 1930's that Kolmogorov succeeded in constructing an axiomatic theory of probability, and only then did probability theory acquire the status of a mathematical science.

Thus, we see that pure mathematics is primarily characterized by the presence of concepts rich in their consequences, i.e., mathematical structures formulated briefly and laconically as a system of postulates. From these postulates conclusions are inferred by means of deductive reasoning which do not depend on any additional explanation of how various statements or postulates can be interpreted in the phenomena of the external world. This point of view began to be distinctly developed in mathematics toward the end of the nineteenth century in the works of Frege and Russell, although a bit earlier Boole had, in the process of creating the algebra of logic, considerably broadened the possibilities of logical deduction and had at the same time made the whole system of judgments much more formal.

The game of chess is often regarded as a model of mathematics (Weyl, 1927) or, if you like, a parody of mathematics. In chess the pieces and the squares on the chessboard are the signs of the system, the rules of the game are the rules of inference, the initial position of the pieces is a system of axioms, and the subsequent positions are the formulae inferred from the axioms. The initial position and rules of the game prove to be very rich; in skillful hands they give rise to a great number of interesting games. Whereas the aim of chess is to checkmate the opponent, the aim of mathematical proof is to obtain a number of theorems. In both cases it is important not just to achieve the aim but to do this elegantly and, certainly, without contradictions. In mathematics certain situations are regarded as contradictory in the same way as, for instance, ten queens of the same color contradict the chess calculus. But the most interesting thing about this comparison is that in chess, as well as in mathematics, logical operations are carried out without any interpretation in terms of the phenomena in the external world. For instance, we do not care what the pawns correspond to in the external world or whether the restrictions imposed on the rules of operating on them are reasonable.

Now we can pass to comparing pure and applied mathematics. But before doing this I would like to mention those limitations which are imposed on the system of deductive constructions. Without this, the reader would get an exaggerated idea of the possibilities of deductive logic.

Limitations Imposed on Deductive Forms of Thinking by Gödel's Theorem

The principle of verification is a criterion of scientific thinking. A statement becomes scientific if it is formulated so that it can be tested. In the natural sciences, the testing of hypotheses is accomplished by comparing them with the behavior of the external world with the help of specific experiments or specifically organized observations (the latter might also be called "experiments"). On the face of it, everything seems quite simple here, but in fact this is not so. The problem of verification, even in the natural sciences, is fraught with great, and in the last analysis probably insurmountable, logical difficulties. (For details, see Chapter 1.) In testing hypotheses we have to face a vexing asymmetry: on the one hand, no experiment supporting the hypothesis is sufficient for its unconditional acceptance, and on the other hand, a single negative result is sufficient for rejecting the hypothesis. A hypothesis in natural sciences is always open to a further test, and this, according to Popper (1965), accounts for the progress of natural sciences. To overcome logical difficulties connected with the acceptance of hypotheses, we have to resort to the language of probability theory when formulating certain rules of the Game against Nature. (This is discussed in greater detail in Chapter 2 of Nalimov, 1971.)

Other difficulties, perhaps more serious ones, are connected with the problem of verification in mathematics. I have already mentioned that in mathematics there is no question of testing the truth of the initial postulates by means of comparing them with observations of the external world. The principle of verification is fulfilled here in the requirement imposed on the system of axioms that they be internally consistent. This means that they must be formulated so that there is no possibility of inferring from them theorems contradicting one another. Correctness of the system of mathematical judgments lies in their consistency.

The need to test the internal consistency of axiomatic systems was a natural consequence of the growing abstractness of mathematical constructions because it was necessary to give a logical foundation to the right to existence of these mathematical structures which determined the development of the separate branches of mathematics. The question of the internal relations between axioms had been troubling mathematicians ever since ancient times, i.e., immediately after the appearance of the Euclidean axioms, the first mathematical structure well known to us. Many efforts had been wasted on attempts to derive the fifth postulate (the postulate of parallels) from the rest of the postulates. But after the non-Euclidean geometries appeared, the question was reformulated, and it became necessary to show their internal consistency. Mathematicians first confined themselves to proofs of relative consistency. The method of mathematical simulation was used. For this purpose it was necessary to construct, within the system of known or generally accepted mathematical structures, models in which the axioms of new structures would be fulfilled. One system of mathematical constructions was interpreted by means of another. When this succeeded, it was stated that the new structures were internally consistent if the old ones were consistent. Thus, for example, a plane in Riemannian geometry is simulated by the surface of a sphere in the three-dimensional Euclidean space. On this sphere arcs of a great circle correspond to the straight lines. Then, indeed, it is impossible to draw through an arbitrarily given point on the surface of the sphere any arc of a great circle which would not cross an arbitrarily chosen circumference of the great circle on the surface of the same sphere. Riemannian postulates thus turn into theorems of Euclidean geometry. The next step was made by Hilbert, who showed that the Euclidean postulates are fulfilled for a certain algebraic model and are consequently consistent if algebra is consistent.

The question of consistency became especially acute after contradictions had been found in the Cantor theory of sets, particularly the famous paradox of Bertrand Russell, which was later named after him. It is noteworthy that Russell, especially in his joint book with Whitehead (Russell and Whitehead, 1910), strives (after Frege) to present pure mathematics as a part of formal logic. Mathematics is now simulated in terms of logic. And if the axioms of arithmetic are only the expression of suitable logical theorems, then the compatibility of the corresponding axioms of logic will follow from the consistency of arithmetic.

The contradictions revealed in the theory of sets, which constitutes the basis of many branches of mathematics, created a crisis situation. In 1904, Hilbert attacked the problem of the absolute consistency of arithmetic, thus acknowledging the insufficiency of proofs of relative consistency. Later, during the decade 1920–1930, Hilbert and his school published a number of papers reporting certain results which, as it seemed at the time, implied the consistency not only of arithmetic but also of the theory of sets. But in 1931, Gödel proved his famous theorem "On Formally Undecidable Propositions of *Principia Mathematica* and Related Systems," which implied the failure of the attempts of Hilbert and his school. (*Principia Mathematica* is the title of the above-mentioned book by Russell and Whitehead.)

There certain logical systems called recursive logics are described. The axioms are regarded as certain "strings" of symbols and the rules on inference are regarded as means of obtaining "strings" out of "strings." Two restrictions are imposed on the rules of inference: they should be finitistic and strictly deterministic. The term "finitistic method" was not strictly defined by Hilbert, and later a heated discussion of its meaning broke out. Evidently, the term should be regarded as defining a method in which transfinite induction is not used. (The latter arises when one has to resort to those transfinite numbers which appear when the notion of an ordinal number is generalized to infinite sets.) The term "deterministic method" need not cause any bewilderment. It means the use of certain unambiguous rules which, when applied to similar data, always yield coinciding results.

The proof of Gödel's theorem on undecidability is too complicated to present here. It is preceded by forty-six preliminary definitions and several auxiliary theorems. A detailed description of Gödel's theorem and its significance for modern mathematics is presented by Kleene (1952). Attempts to give a simple proof of the theorem are made by Nagel and Newman (1960) and by Arbib (1964). I shall only briefly remark that an important role in proving this theorem is played by the arithmetization of mathematics, which is traditionally called Gödel numbering. Every mathematical statement is coded here by an arithmetic formula. The study of mathematical statements is brought down to the study of arithmetic relations.

From Gödel's theorem, it follows that all generally used logical systems in which arithmetic is expressible are incomplete, if they are consistent. There exist true statements expressible in the language of these systems which cannot be proved in the system itself. Furthermore, it follows from the same theorem that it is impossible to prove the consistency of arithmetic systems by means of concepts which can be expressed in this logic. Another consequence from this theorem is that however greatly one increases the number of axioms of this logic, their number being fixed, it will not make the system complete. There will always be new truths which can be expressed in terms of this logic but not inferred from it.[4]

It is difficult to give a formal definition of the concept of "proof" in mathematics. In the process of the development of mathematics, there appear new ways of proving previously unthinkable results. Kleene (1952) expresses this idea as follows:

> . . . we can imagine an omniscient number–theorist. We should expect that this ability to see infinitely many facts at once would enable him to recognize as correct some principle of deduction which we could not discover ourselves. But any correct formal system which he could reveal to us, telling us how it works, without telling us why, would still be incomplete. (p. 303)

Consequently, when mathematics is regarded as a strictly formalized axiomatic–deductive method, the restrictions which are imposed on this statement by Gödel's proof should not be overlooked.

In concluding this section, I would like once more to draw the reader's attention to the fact that the name Hilbert is also connected with the creation of metamathematics, a discipline concerned with the theory of proofs. The object of metamathematics is mathematics, and the latter is spoken of in a language which is a metalanguage with respect to that of mathematics.

And whereas mathematics is a strictly formalized system, metamathematics appears to be intuitively meaningful (though it can be formalized as well), and its statements are formulated in ordinary language. In this connection Kleene (1952) says:

> The assertions of the metatheory must be understood. The deductions must carry conviction. They must proceed by intuitive inferences and not, as the deductions in the formal theory, by applications of stated rules. Rules have been stated to formalize the object theory, but now we must understand without rules how those rules work. An intuitive mathematics is necessary even to define formal mathematics. (p. 62)

In the following section I shall try to make certain metamathematical statements concerning applied mathematics.

---

[4] Later, Gentzen managed to overcome Gödel's difficulty in proving the consistency of arithmetic, but in doing this he had to use transfinite induction up to the ordinal number surpassing all the numbers of an infinite sequence built in a certain way. Such a mode of proof, however, can no longer be regarded as legitimate. I cannot go into particulars in discussing this extremely complicated question and shall not consider all the other attempts to prove the consistency of arithmetic; all of them, in any case, exceed the framework of the problem formulated by Hilbert.

Mosaic Structures of the System of Judgments in Applied Mathematics

Now we can finally pass to the analysis of the system of judgments in applied mathematics. But first of all we should agree as to what we shall mean by applied mathematics. It hardly makes sense to use the term "applied mathematics" for such branches of knowledge as theoretical mechanics or hydrodynamics. They are independent branches of study built axiomatically and deductively, and they have internal logical structures fairly similar to those of pure mathematics in the sense of Bourbaki. It seems to me that it became possible to speak of applied mathematics as a specific and original phenomenon only after the beginning of the process traditionally called "mathematization of knowledge." This is the consequence of the penetration of mathematics into subjects connected with a direct evaluation and interpretation of complex systems of data. Such systems are often called diffuse: no partition can be made which would clearly single out phenomena of the same origin. In such a situation it is difficult to make any precise statement concerning the mechanism underlying the phenomena, and thus it is obviously impossible to construct axiomatic concepts. Later, it became clear that such complex systems should be not only studied and described but also controlled.

For problems of applied mathematics of this kind, the following peculiarity turned out to be typical: integrated mathematical structures

Mathematization of Knowledge

rich in logical consequences disappeared from sight. They were replaced, in some cases, by a multi-colored mosaic of criteria, and for this mosaic structure, the formulation of the question of consistency, which was so significant for the structures of pure mathematics, lost its sense. In other cases, certain statements based on uncertain intuitive considerations were expressed in mathematical language. Here the chain of syllogisms whose presence is at least an outward characteristic of mathematical construction disappeared completely.

Let us try to illustrate this with examples. First of all, consider two examples connected with the construction of logical structures of a mosaic character. The first is a problem of response surface experimental design (Nalimov, 1971). Imagine a sufficiently complicated process dependent on a great number of variables. It may be, for instance, any actual technological process. The mechanism underlying the phenomena of such a process is unknown, and an attempt to study it is rather senseless because of the complexity of the system. It is reasonable to restrict oneself to performing experiments aimed at obtaining a simple (polynomial) model whose geometric image will be a response surface. On the response surface, a region should be found which would ensure the optimal conditions for the technological process. If the possible bounds of variation of every variable are coded with the numbers $+1$ and -1, the region of experimentation may be given by the multidimensional cube with vertices whose coordinates are given by the permutation of the numbers $\pm 1, \pm 1, \ldots, \pm 1$ or by the sphere inscribed within the cube.

The problem of experimental design is to choose in a certain optimal way the location of experimental points in the region of experimentation. If we wanted to solve the problem in a manner strictly corresponding to the general ideas of mathematical statistics, we should require such a location of N experimental points for which the matrix $\mathbf{X}$ of the independent variables would be arranged so that the determinant of the information matrix $\mathbf{X}^x\mathbf{X}/N$ would be maximal on the whole possible set of designs. Such a design is called D-optimal. In this case, the ellipsoid of variance of the regression coefficient estimators will be minimal.

From this standpoint R. Fisher, one of the founders of modern mathematical statistics, developed the concept of algorithm choice for obtaining the best estimators in evaluating experimental data. Here we shall follow the same logical conception, but extend it to solve a broader problem, that of choosing the optimal experimental design. In this case we would seem to act in the frame of the logical constructions which can be called the structure of mathematical statistics, but, in fact, the matter is much more complex. First of all, it turns out that, using the above criteria, it is impossible, strictly speaking, to build an experimental design with a reasonably small number of experimental points. True, this difficulty has been overcome: it has turned out that by using computers it is possible to construct a quasi D-optimal design with the number of points exceeding only by a small margin the number of parameters estimated (see Nalimov and Golikova, 1971). There was another problem which was not so easy to cope with, namely, the necessity of taking into account a number of other optimality criteria which were also in no way connected with the central idea of the development of mathematical statistics. Here, as previously (Nalimov and Golikova, 1971), I shall enumerate all the known criteria:

(1) D-optimality: max $| \mathbf{X}^x\mathbf{X}/N |$.

(2) G-optimality: min max $\sigma^2\{y\}$ where y is the value of the response function.

(3) The minimum of the average variance for the response function: min $\bar{\sigma}^2\{y\}$.

(4) The minimum of the maximal variance of the regression coefficient estimates: min max $\sigma^2\{b_i\}$.

(5) A-optimality: the minimal sum of the squares of the principal semiaxes of the ellipsoid of variance of the regression coefficient estimates (the minimal trace of the covariance matrix $[(\mathbf{X}^x\mathbf{X})/N]^{-1}$ of the regression equation).

(6) The minimum of the major axis of the ellipsoid of variance.

(7) The maximal precision of the estimate of the extremum coordinates.

(8) The minimal error under the condition that the response surface of order $d + 1$ is approximated by the polynomial of degree d.

(9) The best possibility to estimate the lack of the representation of observational results by a polynomial of a given degree.

(10) Proximity of the number of observations to that of parameters estimated.

(11) Rotatability: the matrix $\mathbf{X}^x\mathbf{X}$ is invariant with respect to the orthogonal rotation of the coordinates.

(12) Orthogonality: cov $\{b_i\, b_j\} = 0$.

(13) The possibility of making a definite type of nonlinear transformations of independent variables preserving the design optimality.

(14) The possibility of splitting the design into orthogonal blocks in order to eliminate the uncontrollable time drift.

(15) Composite designs: the possibility of using the points of design constructed to present the results by a polynomial of degree d as a subset of points for an optimal design of degree $d + 1$; this problem may arise when the polynomial of degree d presents the observational results inadequately.

(16) Non-sensitivity to separate flagrant errors in observational results.
(17) Simplicity of computations.
(18) Non-sensitivity to errors in independent variables.
(19) The possibility of estimating, at every step, the lack of fit.
(20) Visual representation of results.
(21) Uniformity: $\sigma^2\{y\}$ should not depend on the distance from the center of the experiment.
(22) The minimal sum of relative errors in regression coefficient estimate.

Not all the criteria are equally significant; some of them, e.g., 17–20, can be regarded as desirable properties of the design rather than independent criteria. We still have about fifteen indisputable criteria. They may be regarded as axioms of designing; the designs corresponding to these axioms may be regarded as theorems. In some cases these theorems result from using comparatively simple mathematical methods, e.g., linear algebra, in constructing a rotatable design. In other cases, e.g., in developing the concept of D-optimality, quite up-to-date mathematics is used: the theory of games, the theory of sets, and functional analysis. Here the system of axioms is obviously of a mosaic character. Of such a mosaic system of initial premises, it is senseless to ask the question concerning its internal consistency. Only in the simplest case, for linear designs, do some of the above-mentioned criteria prove to be compatible. This means that it is possible to build designs satisfying several criteria simultaneously. But everything becomes much worse for the second-order designs, especially when they are given discretely (for further details concerning the compatibility of criteria, see Nalimov and Golikova, 1971). It immediately becomes clear, without any additional analysis, that criteria 8 and 9, for example, are mutually contradictory. The mosaic of the structure of initial premises entails some difficulties (rather unusual for traditional mathematics) in systematizing and classifying designs.

The pragmatic problem is not simply solved either: it is not easy for a statistician to speak with an experimenter. It is possible, of course, to try to appeal to a hypothetical "omniscient experimenter," assuming that in the different concrete situations he will somehow be able to choose one of the criteria. But this does not seem altogether realistic. An experimenter discussing his problem with a mathematician cannot even decide unequivocally the main question: whether the design should be constructed on a multidimensional cube or on a sphere inscribed in it. For a mathematician the sphere is attractive for its natural symmetry. But for an experimenter it is difficult to imagine a region in the space limited by a multidimensional sphere; he gives the bounds of variation of each variable independently, and for him the region of experimentation is limited by a parallelepiped which, after a simple transformation of variables, naturally turns into a cube. The use of vertices of the multidimensional cube enlarges the region of the space under consideration and sharply improves the design characteristics from the standpoint of their D-optimality. However, such an asymmetrical (with respect to the center of the experiment) increase of the region studied may lead to unjustifiable difficulties, e.g., to the necessity of increasing the order of the approximating polynomial.

But now let us assume that we have somehow managed to agree upon the way the restrictions are imposed on the experimental region of the factor space and that the experimenter has informed the mathematician that he is first of all interested in finding the region of response surface where the extremum is situated. A decision has been taken to run a quasi D-optimal design. After the evaluation of experimental results, it has turned out that the extremum is situated somewhere very far out of the design region and, consequently, does not make sense physically. The purpose of the study is then changed: the experimenter says that now he only wants to estimate the contribution made by the separate regression coefficients. This problem may prove unsolvable, since in certain quasi D-optimal designs the regression coefficients are correlated, the correlation coefficient being close to unity.

The situation will become more complicated if we broaden the problem formulation and consider experimental designs aimed at singling out or, as it is usually put, screening out the dominant effect out of a fairly large set of potentially possible ones. It has appeared possible to create the concept of a screening experiment (Nalimov, 1971) and to construct an experimental design with fewer experiments than the number of "suspicious" effects. What is important here is that the estimates obtained as a result of these designs prove to be biased and inefficient.

The whole concept of screening experiments appears to be built in defiance of the main premises of mathematical statistics which I would like to regard as a certain logical structure in the sense of Bourbaki. It turns out that the real problems of mathematical statistics are solved outside the boundaries of this structure. Particularly great difficulties arise in teaching experimental design in all its diversity (Nalimov and Golikova, 1971) when one attempts to explain to an audience the whole mosaic structure of the variety of initial premises, which do not yield to systematization. The students complain that all this "creates a mess in their heads." We are tempted to call such mathematics with mosaic structure vulgar, although some fragments of this mosaic, for instance, the concept of D-optimality, seem quite respectable to mathematicians. Someone would perhaps like to compare the present-day state of things with that existing in geometry. Indeed, as is known, there are many geometries. But, as we have already seen, their axioms can be interpreted on the models of one another. And, what is probably most important, nobody tries to build geometrical constructions which would simultaneously satisfy several axiomatic systems. If, for instance, the general relativity theory has required non-Euclidean geometry, no one will strive to make it Euclidean in addition.

In the field of forecasting stochastic processes, the situation is quite similar, although there are differences. Everything is all right with forecasting stationary stochastic processes. We have here the well-known Kolmogorov–Wiener method, which has been found in the framework of the well-developed system of concepts of stationary stochastic processes. The problem is practically reduced to solving an integral equation whose kernel is the autocorrelation function of the stochastic process. But in fact all or almost all really observable stochastic processes prove to be non-stationary, at least judged by the behavior of their mathematical expectation. There is no mathematical theory of non-stationary stochastic processes. But, as it is put in one paper, there nevertheless exist myriads of publications in which various solutions of the problem are suggested. The best ones have the following structure: a certain model of a stochastic process is suggested which is formulated as an axiom, so that it cannot be either proved or refuted. Proceeding from this model, a formula for forecasting is found by means of constructing a chain of mathematical judgments. In the worst papers the solutions are simply given without even a clear formulation of the initial model. I cannot give here a complete list of the initial model–postulates because nobody has so far tried to classify and codify them. I shall confine myself to considering two well-formulated problems.

One of them is the well-known model of Box and Jenkins (1975). (For its brief description, see Nalimov, 1971.) The principal idea here is to pass from the study of the initial non-stationary process to the study of the sequence of processes created by the first, second, third, etc., differences and to investigate their mutual correlation. It is suggested that one should put down the model of the initial non-stationary process as follows:

$$x_{p+1} = (\gamma_{-l}\Delta^{l-1} + \ldots + \gamma_{-2}\Delta + \gamma_{-1} + \gamma_0 S + \ldots + \gamma_m S^{m+1})\alpha_p + \alpha_{p+1}$$

The letters Δ and S have the following meaning here:

$$\Delta\alpha_p = \alpha_p - \alpha_{p-1}; \ \Delta^l\alpha_p = \Delta(\Delta^{l-1}\alpha_p); \ \Delta^0\alpha_p = \alpha_p;$$
$$S\alpha_p = \sum_{j=0}^{\infty} \alpha_{p-j}; \ S^m\alpha_p = S(S^{m-1}\alpha_p); \ S^0\alpha_p = \alpha_p;$$
$$p = 0; \ \pm 1; \ \ldots; \qquad m \geqslant 1; \ l \geqslant 1;$$

and α_{p+1}; α_p; α_{p-1} are similarly distributed random values with mathematical expectation equal to zero. Having accepted their model, we obtain:

$$\Delta^{m+1}x_{p+1} = (\gamma_{-l}\Delta^{m+l} + \ldots + \gamma_{m-1}\Delta + \gamma_m)\alpha_p + \Delta^{m+1}\alpha_{p+1}$$

$$\Delta^{m+1}x_{p+1} = \sum_{j=0}^{l+m} d_j\alpha_{p-j} + \alpha_{p+1}$$

Hence it follows that all the serial correlation coefficients of order higher than $l + m + 1$ for the differences of the process $\Delta^{m+1}\alpha_p$ of order $m + 1$ should equal zero. If, for instance, all the serial correlation coefficients beginning with the fourth are equal to zero in the sequence created by the second difference, this means that the model with the three parameters γ_{-2}, γ_{-1}, γ_0 can be accepted. Other problems, some of them purely technical, arise here: to estimate these parameters, to correct them with respect to time (since the non-stationarity itself may prove non-stationary), and to evaluate the transfer function determining the inertia the system is subject to.

It is noteworthy that the initial model of a non-stationary stochastic process is formulated as an axiom. Its correctness cannot be tested directly by analyzing the experimental data. Nor can we give any physical interpretation of the model, but from it we receive a working formula allowing us to estimate, on the basis of the experimental results, the number of parameters γ whereby the model for forecasting will be determined. The initial model is an axiom constructed so as to use, in an intuitively obvious way, the correlation properties of the process, although the whole concept is in no way based on the well-developed correlation theory of stationary processes.

The next example is a model suggested by Legostayeva and Shiryayev (1971) for interpolating and extrapolating nonstationary stochastic processes. The latter is given by the model

$$\xi(t) = f(t) + \eta(t), \quad -\infty < t < \infty$$

where $\eta(t)$ is a white noise[5] process with mathematical expectation equal to zero, $f(t)$ is the trend of the process for which it is known that $f(t) \in F_n(M)$ and $F_n(M)$ is a class of real-valued functions $f(t)$ presented as

$$f(t) = a_0 + a_1 t + \ldots + a_n t^n + g(t)t^{n+1}$$

where the coefficients are $|a_i| < \infty$, $t = 0, 1, \ldots, n$ and the measured

---

[5] A random process with a constant spectral density. In other words, this is a noise with no frequency selectivity.

functions $g(t)$ are such that sup $|g(t)| \leqslant M$, M being an unknown constant. On the basis of the results of observations over the stochastic process, it is necessary to find the values of the parameters $a_0, a_1, \ldots, a_n$ in the best way, in a certain sense, by using a specifically determined minimax weight function for which a theorem is proved concerning necessary and sufficient minimax conditions on a certain class of functions. The model is given here in a polynomial form, and for this reason it is possible to make a comparison with the parameter estimates by the maximum likelihood method according to the usual pattern of regression analysis. In comparing maximal variances for the estimate of the parameter a_0, this method has an advantage of the factor 1.25; if the comparison is made in standard deviation having the same dimensions as the values measured, the gain will be only 12 percent; for the rest of the parameters, the gain will evidently be even less. It is important to note that the estimates turn out to be unbiased.

Here some questions naturally arise: in what way can a comparison of two different initial model axioms be made, and in what way are they to be compared with other models? Furthermore, will they yield results differing essentially from the extrapolation in a simple polynomial presentation of the results? The answers to these questions cannot be given in the general case simply for the reason that the models are formulated so as to prove incomparable. None of the models described above can be compared with models in which the forecasting is based on the moving average of an exponentially weighted means, even though all the models are built so as to yield an algorithm for the transformation of data which would give us processes of the "white noise" type. Here again, we are dealing with a mosaic structure of the model axioms.

Now let us consider an applied extremal problem solved outside the framework of an axiomatic–deductive construction. We have in mind a problem of adaptive optimization, namely, the problem of tracing an uncontrollable time trend in a technological process and of adapting it to changing conditions (Nalimov, 1971; Nalimov and Golikova, 1971). In order to solve the problem, a slight "rocking" of the technological process is suggested, i.e., running a permanent experiment directly at the plant, modifying the variables in a small interval in order not to change the technological regime too much. In accordance with this idea, the industrial process will yield not only the necessary production but also the information concerning the direction in which we should move in order to trace the shifting extremum. Having thus formulated the problem, we should worry not only about the experimental design, i.e., the optimal location of the experimental points, but about the whole strategy as well.

Several solutions of the problem have been suggested. One of them is the method of Evolutionary Operation, which consists in locating experimental points according to the scheme of a rotatable design (we spoke of this criterion above on p. 55) and repeating them many times to detect the signal on the background of large industrial interference. The strategy of movement turned out to be unformalized. After a cycle of experiments had been completed, the engineers gathered for a discussion and decided what to do next: to make the steepest ascent, to shift the center of the experiment, and to run a new series of experiments or, most likely, to include new variables in their consideration.

Another method is simplex designing, in which an algorithm for following the shifting extremum is already given. The algorithm consists of the following steps: the experiments are performed in $k + 1$ vertices of a regular k-dimensional simplex with k independent variables; then the vertex with the minimal yield of the process is reflected about the simplex verge opposite to it. The new simplex is then considered consisting of the new reflected vertex and k vertices of the initial simplex.

Then the same procedure is repeated and the minimal yield vertex is reflected again. Several more rules are included in the algorithm, allowing us to avoid "spiralling" around the point with the maximal error of the yield; we shall not consider these details of the algorithm. The attractiveness of the method is obvious: in order to make a decision at each step we use only one new experiment and k previous ones; the greater the number of independent variables k, the more efficient the method. The whole concept is developed outside the framework of an axiomatic–deductive construction. If the formulation of the algorithm is regarded as an axiom, no theorems can be inferred from it, for we cannot regard as a theorem the expression of the coordinates of the mirror reflection of a vertex. True, we can speak of the algorithm in mathematical language. We can say that, if the region of experimentation in the space of independent variables is given by a hypercube and the number of independent variables k satisfies the condition $(k + 1) \equiv 0 \pmod 4$, the experimental design used here is optimal in a broad sense: it is D-optimal, rotatable, and orthogonal. It is also possible to show that, for experiments being performed according to this design, the gradient of the linear approximation obtained will be directed from the center of the simplex to the verge where the reflection is performed. But these can hardly be called theorems, for they are all obvious properties of the given algorithm. The initial premises are not too clear either. The possibility of moving along such a simplex procedure is indubitable if we deal with a linear field; everything will also be in order if the linear field is cut with small ravines, if only the simplex scale is successfully chosen so as to jump over the ravines. But here our conclusions obviously lose their rigor.

Some curious difficulties have arisen in comparing the regular procedure of this kind with random ones. The simplex procedure has immediately been contrasted to the method of random search. In its simplest form the latter reduces to the following: an initial point x_k is chosen in a k-dimensional space; a straight line is drawn through it in a random direction; on this line, on both sides from x_k at the distance p_k two experiments are performed; the experiment with better results determines a new initial point x_{k+1} for a random construction of the second line, etc. Strictly speaking, a random search no longer includes the problem of experimental design; this is a procedure where only the strategy is given. The comparison of the random search method with the simplex procedure can be made only by way of simulating problems on computers. But even this approach requires that the criteria of comparison be chosen and that the conditions of simulating experiments be strictly stipulated. Meshalkin and Nguen (1966) demanded that p_k should equal p_r, the radius of the sphere circumscribed around the simplex. From the standpoint of the mathematician, such an approach proved quite logical; it resulted in constructing a precise mathematical system of judgments, and it appeared possible to prove a number of lemmas and theorems. On the part of the experimenter, however, such a requirement caused perplexity; in performing a random search, the researcher, even in the second experiment, crosses the boundaries of the cube limiting the experimentation region of the space of independent variables. The higher the dimension of the space of independent variables is, the less advantageous the conditions are under which the simplex procedure is placed: it will be performed in the sphere of a smaller radius than that of the random search procedure (see Nalimov, 1966).

In order to make the random search strategy comparable with the simplex procedure, the former should be modified in a special way. Here the very formulation of the problem becomes odd: an algorithm of an applied significance should be modified to become comparable with another one. Rastrigin (1966) gives an interesting collection of criteria for searching out an extremum. It is divided into local and global criteria. In local criteria, losses during searches are considered, i.e., "fast actions" at one step and the probability of an error—the probability of an erroneous step. In the non-local criteria, the number of trials is considered which is necessary for solving the problem set with a given "divergence" (precision) understood as the average deviation of the value found from the extremum in a given situation. Obviously, it does not demand too strong an imagination to increase the number of criteria for comparing two so difficult-to-compare strategies; using these criteria, we shall still obtain new results. Is there any sense in all these activities?

Thus we see that the concept of a simplex procedure does not possess a chain of syllogisms which, as previously mentioned, is at least an apparent feature of the construction of traditional mathematical judgments. Even if we agree to consider the coordinate values of the simplex-reflected vertices as theorems, the results will be fairly poor. The whole concept has a more profound sense which is intuitively clear but cannot be logically inferred from the initial postulate. The attempt made by mathematicians to find this sense by way of comparing the simplex procedure with other procedures was not a success. It is particularly important that it could not have been a success, since such a comparison requires the introduction of new axioms of comparison which determine the results of the comparison and which are in no logical way connected with the formulation of the initial axiom determining the strategy of the method. A variety of mutually uncoordinated axioms of comparison again create a mosaic structure. There is a temptation to call such mathematics vulgar, although it is real in applied problems.

Finally, I would like to answer one more question: Is what is now taking place in applied mathematics a situation similar to that in pure mathematics at the time, for instance, of Newton and Leibniz? Then the concept of mathematical structure did not yet exist. In any case, mathematicians had learned to differentiate before it was really understood what a function is (Box and Jenkins, 1975). It seems that in making such a comparison we have to point out an immense difference. Even at the seventeenth-century stage of their knowledge, mathematicians reached clear and unambiguous decisions although, as a rule, they could not formulate them as theorems. If we are allowed to speak from the standpoint of Platonic realism,[6] it will be possible to formulate a hypothesis that mathematicians acted as if they already surmised the existence of structures undiscovered at the time. In any case, this is how Bourbaki (1960, p. 188) describes the state of things in mathematics of the seventeenth century:

> One must take into account the fact that the way to modern analysis was not opened until Newton or Leibniz, turning their back on the past, undertook a provisional justification of the new methods. They were sustained not by the rigor of methods but by the fecundity of the results.

In applied mathematics or, to be more precise, in the applied mathematics that we are considering, there are a great number of results, but obviously no consistency among them.

---

[6] There are doctrines, shared by some mathematicians (evidently Gödel among others), according to which mathematicians do not create their structures but, like physicists, discover them.

Applied Mathematics as a Language: The Role of Sense Content Underlying the System of Signs

As stated at the beginning of this chapter, the "language of mathematics" is a frequently repeated juxtaposition, a specific linguistic cliché. Let us now subject this phrase to an analysis and attempt to reveal the meaning which it can have.

In many applied problems of a clearly non-mathematical character, the mathematical calculus is used simply as a language which allows us to obtain quickly logical consequences from the initial premises. This language is convenient for its compactness and precision. Since it is well known, there is no need to explain and justify the rules of inference each time anew. Finally, in the process of using this language, which is in a sense a universal language, associations arise with other problems that have been solved with the help of a similar series of judgments, and this gives additional cogency to the new constructions. Mathematics is used here simply as a language for the concise expression of a system of logical judgments. In this connection I would like to remind the reader of the widely known, but far from universally acknowledged, thesis of Frege and Russell that mathematics is merely a part of logic.

When using mathematics as a language, the researcher still does not act as a pure mathematician. He always takes into consideration what lies behind mathematical symbols in various concrete problems. And while the first serious difference between applied mathematics and pure mathematics consists in the absence of a system of judgments of unified logical structures rich in logical consequences in the former, the second difference, of no less importance, consists in the fact that in applied problems of the type in question it is necessary to observe fairly closely what lies behind the symbols. Several examples illustrating this statement are given below.

The first example is as follows. In studying the problems of measuring the growth of science (Nalimov and Mul'chenko, 1969), we formulated an informational model of the development of science. In this model publications are viewed as primary carriers of information. The following postulates were formulated which give the growth of publications in various situations:

$$\frac{dy}{dt} = ky$$

or

$$\frac{dy}{dt} = ky(b - y)$$

where y is the number of publications, k and b are certain constants, and t is time.

The first postulate states that the rate of growth of the number of publications should increase proportionally to their present number. This postulate can be accepted for a situation where there are no factors hampering the process of growth. The second postulate expresses the simplest mechanism of self-braking which begins to take effect only when the number of publications becomes comparable with the constant *b*. By integrating, we obtain in the first case an exponential function and in the second case an equation of an S-shaped logistic curve. These functions are further used to describe the really observable phenomena (function parameters are then, naturally, estimated, the adequacy hypothesis tested, etc.). The curves of growth given by the exponential function can be extrapolated into the future, which drives them to obviously absurd values. This indicates that the mechanism of growth must change. One can consider complicated situations when various countries and different branches of knowledge enter into the game. The observational results should then be presented by a sum of exponents, but this is not too convenient. When one expands the sum of exponents into a Taylor series and confines oneself to the first term, it is possible to limit oneself to presenting results by a varying exponent whose parameters remain constant only in a limited interval of time.

In short, on the basis of the above-mentioned, simply formulated postulates, we obtain rich logical consequences that allow us to discuss easily rather complicated situations. The validity of our judgments increases when we recall that analogous systems of considerations are used in biology to describe the processes of population growth, and in physics for deducing the law of light absorption or the law of radioactive decay. It is pleasant to realize that in all these cases we make use of the same logical structures by operating with the same universal language. But in such a way of reasoning, we always remember what lies behind the symbols and formulas composed of symbols. Imagine the following mental experiment: a set of publications and a portion of radioactive substance are delivered to the Moon. Both the growth of publications and the decay of radioactive substance follow an exponential function. But without any additional tests or reasoning, we shall say that the radioactive substance will continue to decay along the exponential curve while the publications will not grow. When solving the differential equation, we acted as pure mathematicians: we did not trouble about the meaning of the symbols. But when we interpret the resulting functions obtained, we already think about what lies behind the symbols; consequently, we do not think like pure mathematicians.

The second example is an error of interpretation of Zipf's law. Let us assume that there is a certain text with the overall number of words D_N constructed on the basis of a vocabulary containing N individual words. Arrange all N words according to the frequency of their occurrence in the text under consideration. Let the absolute frequency of the nth word (in the order of rarity) in the text be d_n. Then Zipf's law is put down as follows:

$$d_n = \frac{k}{n}$$

where k is a constant determined by the normalizing condition:

$$D_N = d_1 + k \ln N$$

Now let us assume that somebody wants to compute the value of D_{N+1} by using this expression and by substituting $N + 1$ for N under the logarithm sign. Can this be done? If the new $(N + 1)$th word of our vocabulary at the same time takes up the $(N + 1)$th place according to the frequency of its occurrence, this can certainly be done. But now imagine that in the vocabulary there appears a new word, e.g., "cosmonaut." It will not take up the last place in a series of words constructed according to their frequency of occurrence; a rearrangement of words will have to happen, and the parameter will no longer be constant, i.e., a re-normalization will take place. In this case we cannot compute the value D_{N+1} without knowing the new value of k. This specific restriction imposed on the normalizing expression is not expressed mathematically. The researcher should keep this in mind. If he intends to use the normalizing expression as an extrapolating formula, he should think of what places in the list will be taken up by new words. He should think of what is not expressed but only implied: this obviously does not correspond to the mode of thinking of a pure mathematician.

If one does not pay attention to the meaning behind the formulas, one can obtain quite improbable results. I once came across a publication where a normalizing expression analogous to the one mentioned above was used to study the system in development. Regarding N as a function of time, the author began differentiating the normalizing function with respect to time, assuming that the parameter k remains constant. On the basis of the ensuing results, he came to interesting conclusions. When his attention was drawn to the impossibility of doing this, the whole system of judgments collapsed, since there were no data available concerning the behavior of the derivative dk/dt. It would seem that no information about the system in development can be obtained from an expression which does not contain such information, but the author was eager to do so. The mathematization of knowledge often results not only in the use of mathematics but also in the abuse of it. It is sometimes possible for one to be struck by the thesis (which is, however, never formulated explicitly) that by using mathematics to describe observable phenomena it is possible to obtain information which is not present either in the observational results or in the postulates on which the initial models are constructed. I have come across publications where it is mathematically proven that a human being can have only seven levels of abstraction or it is proven that in any field of knowledge half of the publications covers this field of knowledge and the other half covers the adjacent ones. The question is whether it is possible to construct a system of postulates and definitions from which such conclusions could follow. More often than not, these conclusions are connected with the fact that mathematical expressions in applied problems are treated as in pure mathematics, without thinking too much of what lies behind the various expressions clad in mathematical symbols.

The third example is an approximating formula. Can one construct an approximating formula in applied mathematics guided only by the mutual arrangement of experimentally observed points and not bothering about the vaguely formulated subject matter which underlies these observations? I once came across a paper where the curve of growth of the number of scientists in the Soviet Union was approximated in a deliberately complicated manner. The author divided the curve into separate segments, and for each of them, he invented a specific mechanism described by different differential equations. The models thus obtained showed good agreement with the observed data, and, moreover, they were well coordinated. The author was so much carried away by his constructions that he even came to the point of thinking he had inferred his models not from a certain system of postulates but immediately from observational results! What is worth paying attention to here is that the unsmooth behavior of the growth curve is more reasonably explained not by the functioning of a specific complicated and frequently changing mechanism of growth, but rather by the arbitrary character of decision making by the administrative bodies financing the development of science and taking stock of the number of scientists. (The very definition of the term "scientists" and the system of taking stock of scientists change from time to time.) Then the breaks in the behavior of the curve can be described in terms of fluctuation. The decision concerning the choice of an approximating formula has to be made by taking into account considerations not formulated in the language of mathematics.

The fourth example concerns the application of the classical methods of mathematical physics. The heat conduction equation

$$\frac{\partial u}{\partial t} = \frac{\partial^2 u}{\partial x^2}$$

can be solved for $-\infty < x < +\infty$, $-T < t \leqslant 0$, where T is a positive number. If we are given an initial condition of distribution of temperatures at the present moment

$$u(x, 0) = f(x)$$

then the solution describing the distribution of temperatures in the past can be expressed by the integral (John, 1955)

$$u(x, -t) = \int_{-\infty}^{+\infty} k(s, t)f(x + is)ds$$

where
$$k(s, t) = (4\pi t)^{-1/2}e^{-(s^2/4t)}$$

The mathematician asks: How far back does it make sense to search for the distribution of temperatures in the study of space objects such as the moon, using this equation of heat transfer or its generalized form? The answer to the question should be sought with the help of some additional considerations which are again mathematically inexpressible. In solving the problem of determining the limits of a formula's applicability, we use information which it does not contain.

The last example concerns the use of probabilistic arguments in the field of applied research. Here it is fairly easy to formulate the problem in a deliberately nonsensical way. In the well-known British journal *Nature*, the question of the justification of statistical inference was recently discussed quite seriously. The following example was cited: four kings, Georges I, II, III, and IV of the Hanoverian House, died on one and the same day of the week, on Saturday. The probability of the random event is extremely small here: $(1/7)^4 = 1/2,500$. Would not the mathematician come to the conclusion that Saturday is a fatal day for Georges of the Hanoverian House? Of course not. Using some additional considerations he will reformulate the problem (for further detail, see Nalimov, 1971).

An interesting paradox has been formulated by Kendall (1966), a well-known English statistician. It concerns the experiment of tossing a coin. Not only the way the coin falls (heads or tails) is connected with this event but also the character of sound at the moment of the fall, the duration of the fall, and many other phenomena. The probability of the simultaneous occurrence of all these events is insignificantly small, but on the basis of all these considerations, the mathematician does not come to the conclusion that the coin will not fall. He takes into account a number of additional considerations and formulates the problem in another way. By the way, it therefore follows that we should not treat altogether seriously the statement that, in the process of random molecular combination, a live organism cannot appear (see Quastler, 1964). Even if it turns out that in a system of theoretical calculations the probability of a random emergence of life is equal to 10^{-255} or even less, it still seems sufficiently convincing only if the hypothesis as a whole does not arouse any objections proceeding from some other considerations which are much more general but hard to formalize.

Thus we see that, in the applied problems considered above, mathematics functions as a kind of language. In judgments made in this language, we consider relevant not only and not so much the grammar of this language as what we want to say about the subject matter on the basis of considerations stemming from our deeply *intuitive* ideas. Here it is appropriate to recall the school in the foundations of mathematics which is traditionally called intuitionism. It is connected with the names of Weyl (1927) and Heyting (1956). I shall comment only briefly on the complicated conceptions of this school, which is related not only to mathematics but also to the psychology of thinking. According to intuitionist mathematicians, logic is no more significant than a language whose persuasive power is determined by the intuitive clearness and immediate obviousness of every elementary step of reasoning. By now the majority of mathematicians have evidently abandoned such an attempt to establish the foundation of mathematics. I quote Bourbaki[7] (1960) on this point:

> The intuitionist school, the memory of which is doubtless destined to remain only as an historical curiosity, did at least render the service of forcing its adversaries, that is to say, the majority of mathematicians, to make their positions more precise and to be aware more clearly of both the logical and sentimental reasons for their confidence in mathematics. (p. 56)

Interest in these ideas is by no means exhausted since in the applied problems considered here mathematics functions as a language for which the cogency of judgments can be based in the same way in which the intuitionists once wanted to base the system of judgments in pure mathematics. Statements made in the mathematical language in applied problems should first and foremost be intuitively convincing; this is their basis. Here the distinction between pure and applied mathematics is especially precise.

---

[7] The statement of Bourbaki is probably too strongly worded. It should be remembered that many views of the intuitionists are shared by constructivistic mathematicians. Besides, a few mathematicians occupying themselves with the foundations mathematics still stick to the concept of intuitionism.

Language of Mathematics as a Metalanguage: Mathematical Structures as Grammar of This Metalanguage

We speak of a metalanguage when a hierarchical structure of language is considered. Our everyday language is a metalanguage to the "language" of objects surrounding us. In ordinary language, we use the names of objects and not the objects themselves. We speak of mathematics and its logical foundation in mathematical language. The subjects of mathematics are structures and logical deductions from them, expressed in the language of formulas. The subjects of metamathematics are statements about such formal systems. For instance, the statement "arithmetic is consistent" is a statement of metamathematics.

The language of mathematics used to describe applied problems functions as a metalanguage with respect to the language whereby the problems have been previously formulated and discussed. Sometimes statements made in the metalanguage acquire so general a character that this results in the creation of metatheories;[8] what is meant here is already a hierarchical structure of theories. The metatheory evaluates logical consistency of the theories which are hierarchically lower.

This happened, for instance, to mathematical statistics. Its language became a metalanguage with respect to that of various experimental sciences. In the language of mathematical statistics, statements are made about the judgments formulated in these object languages. These statements have acquired such a broad character that a metatheory has been created, the mathematical theory of experiment. Its principal ideas were formulated in detail in an earlier book (Nalimov, 1971). Here I shall briefly repeat the formulations given there: (i) the mathematical theory of experiment has allowed us to formalize precisely the decision-making process in the experimental testing of hypotheses; (ii) it has stipulated the randomization of experimental conditions in order to get rid of the biased estimates obtained in studying the complicated, so-called diffuse systems; (iii) it has formulated clear requirements for the algorithms of information reduction; (iv) it has formulated the concept of a sequential experiment; (v) it has formulated the concept of the optimal use of the space of independent variables.

In the statement of Kleene cited above (p. 52), it is said that the content of mathematics should be understood intuitively; with the help of intuition, we should understand the way the rules of formal mathematics

---

[8] The term "metatheory" appeared after the term "metaphysics," the latter being used for the first time by Andronicus of Rhodes. When classifying Aristotle's works, he introduced the term "metaphysics" in order to place the philosophical works by Aristotle on the first causes after the works on physics. The Greek word μετα has the meaning "after," behind."

function. In applied problems, mathematics itself functions as a metatheory, and for this reason it should also be intuitively grounded despite the apparently formal language in which it is expressed.

If mathematics plays the role of a language in applied problems, the mathematical structures of this language are naturally considered as the grammar of this language. The question can be asked whether it is necessary for the person whose use of the language will be purely pragmatic to know the grammar perfectly. The answer seems to be negative; in any case, it is possible to speak everyday language without knowing its grammar. I recall that in Russia during the first decade after the revolution of 1917, there existed rather a strange viewpoint according to which it was not necessary to teach grammar in the secondary schools. In fact, it was not taught, but the people graduating from school still had a full mastery of the language.

Above (see p. 64), I gave an example showing the way the language of differential calculus is used to discuss the problems of measuring the growth of science. Is it necessary that the participants in this discussion understand the foundations of calculus based on the concept of set theory? It would seem not: it suffices to have only a general understanding of the rules for differentiation and integration which would be quite similar to these existing in the time of Newton, Leibniz, and their closest followers.

I have already mentioned the fact that probability theory acquired the status of a modern mathematical subject only after Kolmogorov had given it an axiomatic structure. It turned out that probability theory could be constructed in the framework of the general theory of measure with a special assumption that the measure of the whole space should equal unity. (Probability can never exceed unity; this is the maximal probability of the certain event.) The probability theory formulated as a mathematical subject proved to be a part of a very general mathematical concept with a clear logical structure of an absolutely abstract character. But such an approach to the definition of probability appeared to be practically inconceivable for experimenters. It was the frequency theory of probability that had a strong effect upon them; according to it, probability is defined as the limit of the relative frequency of an event when the number of trials increases to infinity. From the viewpoint of the experimenter, this definition seems intuitively obvious though it is logically inconsistent. Kolmogorov (1956) writes that a definition of this kind ". . . would correspond to the definition of a point in geometry as the result of splitting a physical body an infinite number of times, halving its diameter every time." He noted further that a frequency theory of probability which involves such a passage to a limit is in reality a mathematical fiction since it is impossible to imagine an infinite sequence of trials where all the conditions would remain constant. True, he also notes that the solution of applied problems does not necessarily demand a formal definition of probability. It suffices here to speak of probability as a number around which relative frequencies are clustered under specifically formulated conditions, so that this tendency to cluster is manifested more and more accurately and precisely with the growth (up to a reasonable limit) of the number of tests. It is interesting that neither of the two definitions solves the paradoxes (see p. 68) that arise when one tries to apply probabilistic concepts to the description of real problems in a very formal way.

Note also that the grammar of the language of mathematics cannot always be used to construct a system of inferences for real problems. Two examples illustrate this.

In mathematical statistics, a theorem is proved stating that the regression coefficient estimates obtained in problems of a multidimensional regression analysis by the method of least squares prove to be unbiased and efficient in the class of all linear estimates. Generally speaking, this is true only if all independent variables and all the corresponding regression coefficients with the mathematical expectation different from zero are included in the consideration. But mathematicians never stipulate this condition, and they do not have to do so. The mathematician always deals only with the model which is expressed in mathematical symbols. He cannot take into account what is implied but not explicitly expressed. The experimenter's mode of thinking is different. Applying regression analysis, e.g., to describe a technical process in a plant, he is quite aware of the fact that far from all possible and really existing independent variables are included in the mathematical model. Many of them prove not to be included simply for the reason that it is practically impossible to measure them. In this case, the regression coefficient estimates will be biased. This bias can be so large that the results of regression analysis lose any sense. (An example illustrating this statement is analyzed in detail in Nalimov, 1971, p. 162.)

Another example is information reduction. This is a purely linguistic problem. Information contained in a long series of observations over a random variable must be expressed in a compact form. It was suggested that the grammar of these statements be built so that the parameter estimates whereby the statements are generated are unbiased, consistent, and efficient. For a long period of time, all the textbooks of mathematical statistics were full of such recommendations. But later the approach turned out to be too dogmatic. In real problems it is necessary to take into account one more property, namely, robustness, i.e., nonsensitivity to discrepancies from the initial premises concerning the distribution functions. We are more often than not dealing with a contaminated sample where observations taken from one universe mingle with those from another universe with different parameters. Estimates efficient with respect to uncontaminated samples can prove very poor with respect to contaminated samples. We have to abandon the rules of grammar inferred deductively and replace them with recommendations obtained as a result of simulating the problems on computers (for details, see Nalimov, 1971).

Variety of "Dialects" of the Metalanguage of Mathematics

It is well known that a single practical problem can often be expressed and discussed in a variety of mathematical "dialects." Sometimes it can be formulated at the level of deterministic ideas, the hypothetical mechanism governing the process being described with the help of differential equations. At another time the same problem may be discussed in probabilistic terms, and here again various dialects can be used: we speak in terms of classical mathematical statistics on one occasion and in terms of information theory on another. Let us assume, for instance, that we are discussing optimization of a certain technological process. We can try to give a strictly deterministic model of it. Then the problem of optimization will be reduced to an application of the calculus of variations to such new branches of study as the method of dynamic programming and the maximum principle of Pontryagin. If, however, we estimate the level of our knowledge of the mechanism of the process under consideration pessimistically enough, we shall have to confine ourselves to the use of the language of multidimensional regression analysis or of the language of the method of principal components and, perhaps, of that of factor analysis. If anyone finds the probabilistic language still very unpleasant, he may use Boolean algebra. In this case the intervals of variation of independent and dependent variables should be split into separate regions and coded in the binary number system. It is then possible to apply the method of minimizing Boolean functions in the algebra of logic. In this model the target function and predicates will be connected by the logical operators "and" and "or" (for a detailed description of a technological situation, see Shcheglov, 1972).

In a single seminar, one and the same problem can be discussed in a variety of dialects, whereas for ordinary language such a situation is very rare. An adequate translation from one mathematical dialect into another appears to be impossible, just as translation is, strictly speaking, impossible both for ordinary languages and for an abstract, completely formalized language.[9]

It is impossible to give a criterion which would allow us to prefer a certain mathematical dialect in describing a practical problem. Moreover, it is impossible even to suggest a criterion for testing the hypothesis that a certain dialect of the language of mathematics is acceptable for describing a certain situation. It might seem that the following statement could be chosen as such a criterion: the language is accepted for describing a practical problem if by means of it a mathematical model can be obtained which adequately describes the observed phenomenon. But here we may recall one of Russell's paradoxes (see Russell, 1961): assume that somebody regularly hires a taxi and constructs a graph marking the number of the day on the abscissa and the number of the taxi on the ordinate. If n observations are obtained, they can be represented by a polynomial of the degree $(n - 1)$. The curve corresponding to the polynomial will pass through all the points observed. The model will be adequate in some respects, though there do not remain any degrees of freedom to test its adequacy in the familiar statistical sense. However, it is inapplicable for predicting the number of the taxi that will be hired tomorrow.

The same experimental data could be presented as a stochastic process, in which case the problem of forecasting would acquire sense. The question of the choice of a model and consequently of the choice of a dialect cannot be solved by a mere test of the adequacy of the hypothesis. The same difficulty can arise in the problem of interpolation. I once saw a case where, under the experimental conditions, the experimenter could only receive experimental points situated on the left and on the right of a two-dimensional graph, the middle of the graph remaining empty. An approximating formula had to be found that would describe the behavior of the function in the region missing from the graph.

A mathematician immediately, and quite naturally, suggested approximating observational results by a higher-order polynomial. The multiextremal character of the graph of this function aroused the the indignation of the experimenters. This kind of conflict situation where the researcher–experimenter has, on the intuitive level, certain prior information of the mechanism of the process under study but cannot formulate it in a form acceptable to the mathematician is, in fact, very common.

The mathematization of knowledge which has begun recently leads to the appearance of many papers in which identical, or at least similar,

---

[9] It is interesting to note the following: in the abstract mathematical theory of context-free languages it is stated that the problem of finding the finite transformation reflecting the language generated by one context-free grammar to the language generated by another grammar is algorithmically unsolvable (see, e.g., Ginsburg, 1966).

situations are described by a variety of models formulated in different mathematical dialects. Broad application of mathematics only increases the Babelian difficulties in science. Will there appear a criterion that will restrain this process? At present it is hard to answer this question. Such a criterion could be the requirement that we consider legitimate only the use of those dialects of mathematics whose application leads to the creation of useful metatheories. An example of such a metatheory is the mathematical theory of experiment which came into being as a result of the broad application of probabilistic language to the description of experimental situations. But here another question emerges immediately: What is to be considered a useful metatheory? There may appear self-contained metatheories. In the process of creating a metatheory or a fragment of one, a researcher brought up on the tradition of pure mathematics may formulate postulates without caring much about their logical consequences. He may not be a bit worried about whether his logical constructions are realistic.

Polysemy of the Language of Mathematics

The development of the scientific era in which we are living began with a struggle against the polysemy of our language. The Cartesian school of philosophy demanded that scientific terms should be strictly defined. This demand undoubtedly had a favorable effect on the development of the exact and natural sciences. Yet, researchers did not apprehend it completely, and a strictly monosemantic language was not created. Quite recently, the English school of linguistic philosophy put forward the thesis that a natural language is always richer, as a result of its polymorphism,[10] than an artificial language with strictly defined terms (see Gellner, 1959). Earlier (Nalimov and Mul'chenko, 1972), we attempted to strengthen this statement on the basis of Gödel's theorem. From the theorem it follows that human thinking is richer than its deductive formulation. Communication between people takes place at a logical level where strictness is in some polite way broken by the polymorphism of the language, and Gödel's difficulty is thus overcome.

For a long time, the language of mathematics remained strictly monosemantic. It was used only to describe those well-organized systems which were dealt with in traditional physics. Lately, the language of mathematics has begun to be used for describing poorly organized, diffuse systems. In this process it has acquired certain features of polymor-

---

[10] What is meant here is the diversity of meanings ascribed to a sign within the limits of a language dialect. The term "polysemy" might be better, but Gellner (1959) used the term "polymorphism."

phism. The requirements placed on mathematical descriptions have become less strict. Whereas the description of real phenomena in mathematical language was earlier regarded as the expression of a law of nature, now it has become possible to speak of mathematical models. A single system under study can be described by a variety of mathematical models which may all simultaneously be legitimate. This question was considered in detail in Chapter 1 of *Teoriya Eksperimenta* (Nalimov, 1971). Further, we showed (Nalimov and Mul'chenko, 1972) that polymorphism can be observed within the limits of one model. This occurs in the problem of transforming variables in regression analysis where the parameters of transformation can be chosen arbitrarily from a broad range of possible values. In the problem of the spectral presentation of a stochastic process, the experimenter receives not one curve but a variety of curves of spectral density computed on the basis of different weighting functions—the so-called "spectral windows." The mathematician has no basis for choosing among these functions, which are constructed in such a way that an increase in precision of the estimate of the spectrum leads to an increasing bias.

This polymorphism of the language of applied mathematics increases its flexibility. The boundary between it and ordinary language becomes in a sense less obvious, and at the same time, a new difference from traditional mathematics appears. I could also speak of the unpleasant manifestations of the polysemy of mathematical language which arise in connection with solving problems of an applied character. Let us return to the fourth example analyzed above (pp. 67–68).

In order to estimate the temperature distribution in the past, we must know the initial conditions $u(x, 0) = f(x)$. In real problems we can deal only with an approximated sample estimate $\hat{f}(x) \to f(x)$. It turns out that small arbitrary changes in $f(x)$ and in a finite number of its derivatives may lead to great changes in $u(x, -t)$. The problem of temperature distribution for earlier values of time proves incorrect in the same way in which Hadamard formulated his criticism vis-à-vis the problem of Cauchy (the problem of Cauchy consists in finding a solution of a differential equation which satisfies the given initial conditions). Hadamard showed that the formulation of this problem is incorrect for elliptic equations because their solution does not depend continuously on the initial conditions. Here I shall not dwell on the question of incorrectly formulated problems; it has been considered in detail by other authors (e.g., Ivanov, 1963). The search for a correctly formulated problem is a struggle against the troublesome polysemy of mathematical language. Even if one has managed to formulate a problem correctly, it does not yet mean much. For instance, a correct numerical solution of the heat-transfer problem for the past (see John, 1955) does not remove the question of how far back it makes sense to compute it. Russell's paradox mentioned above concerning the prediction of the taxi number springs up in spite of the use of a correct (in this sense) problem formulation.

Mathematical Model as a Question to Nature Asked by a Researcher[11]

The application of mathematical language in research allowed us to formalize our ideas as to what a good experiment is. That gave birth to the mathematical theory of experimental design, which was discussed in an earlier section.

The principal idea of my earlier book (Nalimov, 1971) is that the proper design of experiments is possible only when a mathematical model of the process under investigation has been found; moreover, it is stated there that the possibility of planning and its efficiency depend entirely on how the model is stated. However, my book does not provide any information as to how the models are to be constructed. The reason for this is quite simple: *the construction of a mathematical model is an art, whereas the design of an experiment is mainly a procedure.* Obviously, it is much easier to discuss procedures.

Nevertheless, a few words can be said even on the subject of art. A *mathematical model*, in the sense used here, is a *question* put by the researcher to nature. What, then, are the semantics of a question. The subject of interrogative (erotetic) logic is being given much attention [see, for example, Hintikka (1972) and Kondakov (1971)].

Any question necessarily consists of two component parts: an assertive part which introduces some knowledge, thus making the question possible (this part can be regarded as a prerequisite of the question), and the interrogative part proper. The interrogative part can be neither true nor false; it can only be either relevant or irrelevant. The prerequisite of a question can be true, false, or inadequate for asking the given question, and this can be established only by introducing some other, external information not contained in the question proper. Let us consider two examples.

Example 1. A psychiatrist says to a boy patient, "An American rooster lays an egg on the territory of the USSR; to what country does the egg belong?" The boy answers with astonishment, "How can a rooster . . ."

[11] This section is a part of the booklet written together with T. I. Golikova, *Experimental Design Theory: the Achieved and the Expected*, and was published in the journal *Industrial Laboratory*, No. 10, 1977. The journal is translated into English in the United States, and I have borrowed the translated excerpt from it.

Such an answer satisfies the psychiatrist; the patient perceived the fallibility of the assertive part of the question.

Example 2. At philosophical conferences one is frequently asked the question: "Do you believe that absolute truth exists?" A prerequisite of this question is the implied assertion that there exists (or at least can exist) a language semantically rich enough to express absolute truth. (If such a language cannot exist in principle, what is the difference between the last assertion and the assertion that negates the existence of absolute truth?) However, if one recalls Gödel's theorem,[12] the implied prerequisite of the question seems rather doubtful. We thus see how a question gives rise to a prior question that must be answered before the main question becomes relevant.

Thus, a question can be neither false nor true and, consequently, is no assertion in the strict sense (Kondakov, 1971). Langer (1951), following Cohen (1929), says that a question is an ambiguous sentence whose determinant is the answer to it. Hence it is clear that a question can be irrelevant or even forbidden. The development of any culture is prescribed by a set of permissible and forbidden questions. In our culture, for example, forbidden questions are ones such as "Why does Ohm's law exist?" or "Wherefrom and when did Ohm's law appear?" (see Chapter 1) since any possible answer to these questions seems an absurdity.

Any science, if we speak in contemporary terms, begins with formulation of questions. Generally speaking, one can carry out observations and even conduct experiments without questions, but such activities can hardly be called scientific. Ethnographers tell us that among peoples with a pantheistic outlook one can meet observers who know all that can be seen about nature. However, they observe without questioning, and the reason they do not ask questions is that they have no theory for making the prerequisite of a question meaningful. Alchemy existed for nearly 2,000 years, and its goals resembled those of modern chemistry; in the process of its (rather slow) development, equipment used up to our time was devised, and a lot of discoveries were made. But alchemy was not a science. Experiments were conducted blindly with the single purpose of making gold; no questions were asked about nature, and in fact they could not be asked since there was no language for formulating theoretical concepts based on past experience. But even today, listening to reports or a defense of a thesis, one often feels inclined to ask: "What question is your paper answering?" More often than not the answer is: "Why should it answer a question? That has not been our aim. We have just made this and that . . ."

---

[12] Gnoseological problems generated by Gödel's theorem have been well treated by Nagel and Newman (1960).

A characteristic feature of contemporary science is thus still the fact that scientists aim at getting answers to precisely formulated questions based on previous knowledge. However, knowledge is always relative and changeable since science progresses dialectically, i.e., in a revolutionary manner (Kuhn, 1970a; see also pp. 9–14). Consequently, a more careful formulation would be that scientists ask questions about nature on the basis of contemporary prejudices.

Mathematics is a language in which questions can be asked in a surprisingly compact form with the aid of abstract symbolic notations. Let us assume that a researcher has formulated his problem in the form of a model

$$\eta = \alpha(x\theta)$$

in which he wants experimentally to evaluate the vector of parameters θ. In such a formulation, the model written above is simply a well-asked question. Its prerequisite is a distinct separation between the dependent and independent variables responsible for the investigated process and the analytic description of the model proper; the interrogative part is the specification of the parameter vector which has to be numerically evaluated. In the case of insufficient a priori information, the prerequisite is weakened, and instead of a single model one can have several competitive models, or in place of a single small set of independent variables one has a multitude of variables from which one must select really significant ones by means of a screening experiment. A change in the prerequisite of the question causes an immediate change in its interrogative component.

Earlier (Nalimov, 1971), attention was drawn to the fact that even the most simple problem, such as weighing three objects with the aid of a balance, can be subject to planning provided a model is specified. Such a model must necessarily contain an assertive part, which in the case of weighing is a polynomial model without interaction members (the experimenter asserts on the basis of a priori information that no interaction takes place, and this is the prerequisite of the question). The investigator may have no a priori knowledge about the mechanism of the investigated phenomenon, but still the question may be asked on the basis of some information about the logical structure of some "blind" examination. As an example, one can use the model of staging a screening experiment in pharmacological isolation of therapeutically active or toxic preparations. [This is discussed in detail elsewhere (Nalimov and Golikova, 1976, Section 3 of Chapter 6); the reader can readily separate in the model its assertive and interrogative parts].

One can speak about a hierarchy of interrogative components associated with the mathematical model as a question. If we get an answer to the first interrogative component of the above model, i.e., the numerical estimates of the parameter vector θ are found, we immediately have a second, hierarchically higher, interrogative component: it is necessary to evaluate how the given model describes the problem (to test its adequacy, etc.); in this case the evaluation of parameters turns into the assertive component of the new question. If the model is found to be satisfactory, the information is included in the assertive part of the question, and a new interrogative component appears which otherwise can be formulated as follows: Where is the extremum? What is the response surface in the extremum region?

Any scientific hypothesis, especially when written in a mathematical form, can perhaps be regarded as a question. A probabilistic model of the semantics of everyday language using the Bayesian theorem has been proposed (Nalimov, 1974b, 1981).

$$p(\mu|y) = kp(\mu)p(y|\mu)$$

where $p(\mu)$ is the distribution function of the meaning of word μ given a priori; $p(y|\mu)$ is the likelihood function which defines the distribution of the semantic content of sentence y, provided our attention is drawn to the meaning of word μ; and $p(\mu|y)$ is the a posteriori probability that defines the distribution of the meaning of word μ in sentence y. This model will be discussed in detail in the subsequent chapters of this book. In a profound sense, the prerequisite of this model, if treated as a question, is the assertion that our language is discrete while our thinking is continuous (Nalimov, 1979). The interrogative component of the question is aimed at clarifying how on the basis of the written model one can explain the entire diversity of our verbal behavior.

It should certainly be possible to devise a meaningful classification of models by treating them as questions. But we are still not prepared for this. We shall restrict ourselves here to several brief remarks. One of the noteworthy classes of models is classification models. In a natural way they can be divided into models of logical classifications (for more detail, see Meyen and Shreider, 1976), such as the universal decimal classification used in libraries and models of numerical taxonomy (the method of principal components, cluster analysis, etc.). A characteristic feature of such models is that they describe the observed phenomena regardless of cause-and-effect relations. Numerical taxonomy models are gratifying in their "poverty." Their prerequisites are extremely limited: the observable variables cannot be divided into dependent and independent; the whole thing is limited to enumeration and defining a metric; a stopping rule is selected (in some models) for ceasing the classification procedure;

and the interrogative part is free from many claims and is limited to a search for the hierarchy of taxons in the given imperative metric.

It is interesting to note also that probabilistic models raise questions about the diffuse–behavioral descriptions of the universe (Gellner, 1959) regardless of cause-and-effect relations (For more details, see the next chapter). Differential equations appear any time a question is asked in a cause-and-effect formulation. Of special interest is the recent attempt to use digital computers to construct models with a weak assertive part and an interrogative part full of pretensions. As an example, we cite the grandiose American program for a comprehensive study of five ecosystems: the tundra, steppes, deserts, and leaf-bearing and coniferous forests. Comprehensive models have been constructed and partitioned into blocks containing up to a thousand parameters under very weak initial theoretical prerequisites. The models were to answer questions about the behavior of these ecosystems. It is still too early to speak about the outcome of this program. However, a critical analysis of this activity, based on a careful analysis of materials concerning three of the above-mentioned ecosystems, has been published recently (Mitchell et al., 1976).

It might, of course, be possible to formulate some criteria for good models. But once again we must limit ourselves to certain fragmentary remarks. Levins (1966) states that in models there is a trade-off between *generality, precision*, and *realism*. (Nothing similar is observed in the "laws of nature," and in this they differ from models describing diffuse systems.) Stressing of one of these three factors immediately weakens the others. We could add to this that, with a given assertive part and certain fixed experimental possibilities, the answer is the more definite the weaker the demands specified by the interrogative part of the model. Hence, it can so happen that in the description of chemical processes simple polynomial models can provide more than models with nonlinear parameters which pretend to give an adequate description of the mechanism of the process. (This very important problem will be raised again in Chapter 8.)

Specialists in experimental design seem now to be somewhat disillusioned. All was well when the topic was the design of so-called extremal experiments. All was clear: the model proposed by Box and Wilson in 1961 proved to be typical[13] of many situations, especially in technical sciences. Certain additional typical models, such as screening problem models, became known later. Effort had to be devoted not only to the selection and construction of the model (although the activity still was

---

[13] One possible reason that mathematical statistics is not very much appreciated in science is that many professional statisticians tend to reduce the entire diversity of real problems to certain typical models such as the models of the analysis of variance.

creative[14]) but mostly to the subsequent purely technical part of research activity. Many experimenters have learned how to do this by themselves without outside help. Specialists in experimental design should now become modelers, i.e., constructors of models. Herein probably lies the success both of their personal activities and of the field as a whole.

Is it possible to teach modeling, which is said to be more of an art than a science, and if so, what should be taught? To write a deeply intimate letter to one's friend is also an art, but nevertheless school children are taught writing even if it is known that not all will grasp the art. To educate model builders, one must teach them how to represent in a compact symbolic form assertions about the real world which are posed in a vague, uncertain manner. The model must clarify a specified problem or question. It must exhibit an economy of ends and means. It is an art.

Peculiarities of Teaching Applied Mathematics

The pragmatic meaning of the distinction between pure and applied mathematics first becomes obvious in the problems of teaching. If the logical structure of applied mathematics is different from that of pure mathematics, it should also be taught differently.

One should not think that in speaking of the peculiarities of applied mathematics I want to deny the role of mathematical structures or abstract mathematical constructions. But they seem to reflect a certain outlook rather than serve as a real instrument. Not everybody engaged in studying applied problems should possess this outlook or possess it to an equal extent. In any case, it would have been quite unrealistic to think that every experimental researcher could also become a mathematician. It is also unrealistic to believe that every mathematician will be able to solve applied problems using mathematics of a form foreign to him. Here again, I would like to raise the question of training scientists of an intermediate type (for details, see *On Teaching Mathematical Statistics to Researchers*, 1971). A scientist with a solid background in pure mathematics can prove to be quite helpless in applied mathematics. [After this section had been completed, my attention was drawn to a very interesting article by I. I. Blekhman, A. D. Myshkis, and Ya. G. Panovko (1976) devoted to the same problem.]

---

[14] Creative activity here means the selection of one out of several typical models, the selection of dependent and independent variables, the choice of the region of the independent-variable space in which the experiment is to be carried out, etc. It can be easily calculated that if the error in the selection of the span of variation of each independent variable is 15 percent, then, for example, in the estimation of second-order polynomial models for five factors, the efficiency of a D-test plan decreases by approximately 60 percent as referred to one parameter.

Concluding Remarks

I now return once more to the question of where the demarcation line lies between pure and applied mathematics. Naturally, there can be no precise boundary between them. We may think of a continuous scale of logical structures of various degrees of rigor. At one end is located pure mathematics in the sense of Bourbaki; at the other end, there are those of its applications which I consider related to the broadly developing mathematization of knowledge. The difference between pure and applied mathematics is most prominent when we contrast the extreme manifestations of what is essentially the same phenomenon. Theoretical mechanics and hydrodynamics get quite close to pure mathematics on this scale, and physics obviously occupies an intermediate position. Hutten (1956) made a very interesting statement concerning the logical structure of physics. He said that, in attempting to reconstruct any branch of knowledge, we must distinguish between three stages of formalization: mathematization, when mathematics is used merely as a language; axiomatization; and the construction of interpretation rules. Further, he remarked that if one looks at physics from this standpoint he will have to acknowledge that its formalization has been limited to the first stage, mathematization. Many attempts have been made to axiomatize physical theories, but only one of them, that of Carathéodory, an expert in mathematical analysis of thermodynamics, has gained general recognition. Even the axiomatization of mechanics by Newton has not been a success. As to the interpretation rules (translation from the mathematical language to the experimental one), they are, strictly speaking, absent from physics. A brilliant example is a report by Abel (1969), who presents a collection of diverse and uncoordinated statements of physicists and philosophers, students of the foundations of physics, concerning the interpretation (in the language of experimental physics) of the concept of probability waves (psi-waves). Abel begins his article with the question: "Can we be said to know something . . . which we have not been able to put into words?"

I believe that an opposition of the two extreme manifestations of one and the same phenomenon has its raison d'être at least in the fact that it serves to draw attention to that end of the logical scale where, under the guise of mathematizing knowlededge, something is being done which is very far from what mathematics proper is.

Chapter 4

Why Do We Use Probabilistic Concepts to Describe the World?[1]

Introduction

For a long time science had been developing so that scholars were satisfied with its results if they led to explanations of Nature or to its control (as in the natural sciences) or to inwardly consistent results (as was the case with mathematics). However, criticism began to arise as a result of the drive for self-analysis. Mathematics was the first scientific subject which already at the end of the last century manifested the need to construct its own foundations. The landmarks along this course are the discovery of contradictions in the theory of sets by Russell, Hilbert's intention to prove the absolute consistency of mathematical structures, ending in Gödel's theorem on undecidability, and, finally, metamathematics in its present-day state.

In the natural sciences the necessity to create their foundations was realized much later — this seems to have happened no earlier than in the 1930's. However, now we are able to speak about a fairly distinct and solid trend called "the philosophy of science."

Research into the logic of the development of science has not brought such powerful results as were obtained in metamathematics, but the papers by Carnap, Reichenbach, Popper, Kuhn, and Feyerabend are very interesting. In any case, they make us think about what we are doing in science. At the closing session of the Fifteenth International Congress of Philosophy in Varna (1973), we heard roughly the same: ". . . certain-

[1] This chapter was published in Russian as a preprint, "Language of Probabilistic Notions," by the Scientific Council of Cybernetics, Moscow, 1976. The major part of it was published in the journal *Automatika* (no. 1, 1979), which is translated into English. This chapter was translated by A. V. Yarkho.

Use of Probabilistic Concepts in Descriptions of the World

ly, philosophy has not moved humanity closer to the truth, but it has made it now more difficult to make mistakes."

One of the profound problems of the logic of science is why we use probabilistic concepts to describe the world, in contrast to the traditional deterministic mode of description. Have we any grounds to do this, and if so, what are they? I felt a desire to discuss these questions after reading the book *Theories of Probability. An Examination of Foundations* by Fine (1973) and the critical remarks on it by Tutubalin (1974).

Criticism by T. R. Fine and V. N. Tutubalin

The book by Fine is immediately interesting as a result of his use in the title of *theories*, rather than theory, of probability. Breaking the existing paradigm, he made an attempt to give a mutual comparison of all or almost all existing conceptions of probability, excluding only the rather incomprehensible fiducial probabilities of Fisher, non-commutative[2]

---

[2] This not widely known trend is developed in connection with certain problems of quantum mechanics, where we have to assume the existence of random values with deliberately large variance. For details, see the article by Parthasarathy (1970).

probability theory, and probabilistic concepts as model constructs in human language. The resulting book contains: the classical theory of Laplace, the frequency conceptions of von Mises, the frequency theory of Reichenbach–Salmon, Kolmogorov's axiomatics, the axiomatics of comparative probabilities (Fine's field of interest),[3] the algorithmic approach of Chaitin, Kolmogorov, and Solomonoff to evaluating randomness as complexity, the logical probability of Carnap, and the subjective (or personal) probabilities of Savage and Finetti. The variety of probability theories does not, strictly speaking, form a unified mathematical structure (in terms of Bourbaki). Rather, we observe a mosaic of separate logical structures, which have proved to be interrelated, to some extent. The author has managed to illustrate these interrelations explicitly. More than that, he has attempted, and not unsuccessfully, to demonstrate that each of these theories reflects essential features that we connect with the concept of probability. That leaves us small hope that a general all-embracing theory of probability can be constructed now. At the same time, while reading the book, one cannot help asking why the present-day tradition does not reflect the variety of probabilistic concepts in the textbooks? Even the algorithmic approach to a definition of randomness and probability is lacking. I do not really believe that many specialists in probability theory ever took note of the fact that logical probability proves compatible with frequency-based probability and, moreover, seems to be of great help in elucidating the classical probability concept.

However, I have to acknowledge that if a course of lectures embracing all the theories of probability were presented, the beauty of structures would immediately perish: the architecture of the course would become strange for a mathematican.

Still, the burden of Fine's book is not an attempt to elucidate connections among separate theories of probability, but rather to answer the question of why we use probabilistic concepts to describe the world. Do any theories of probability make this practice legitimate? Do theories of probability possess sufficient grounds for this? Fine answers this question in the negative. Here are some excerpts from his concluding remarks:

> *Finite Relative-Frequency Interpretation of Probability.* The advantage of a finite relative-frequency interpretation of probability is that it easily answers the measurement question, at least for those random phenomena that are unlinkedly repeatable. . . . Finite-relative frequency is descriptive of the past behavior of an experi-

---

[3] This is another trend that is not broadly known. It deals with the approach when the probability of an event does not take a numerical value but we can say about certain pairs of events that one of them is more probable than the other.

ment but it is difficult to justify as being predictive of future behavior. . . . Furthermore, not all random phenomena are amenable to analysis in terms of arbitrary repetitions. Why are not unique occurrences fit for probabilistic analysis? If we look closely, we see that we never exactly repeat any experiment. Nevertheless, it appears that informal recognitions of approximate repetitions coupled with a finite relative-frequency interpretation is the most commonly applied theory of probability.

Limit Relative-Frequency Interpretation. A limit statement without rates of convergence is an idealization that is unlike most of the idealizations in science. . . . Knowing the value of the limit without knowing how it is approached does not assist us in arriving at inferences. The relative-frequency-based theories are inadequate characterizations of chance.

Algorithmic Theory. The theory of computational-complexity-based probability, . . . while successful at categorizing sequences, is as yet insufficiently developed with respect to the concept of probability. However, the indications are that when developed, this approach will be able to measure probability, but will encounter difficulties with the justification of the use of the measured probabilities. The justification problem seems to be very similar to that faced by other logical theories of probability.

Classical Probability. Classical probability . . . is ambiguous as to the grounds for and methods of assessment of probability. It partakes of elements of the logical and subjective concepts and is far less clear than the logical theory as to how to reach probability assessments. It is also perhaps true that the subjectivist claim to subsume classical probability as a special case is valid. In the absence of a clear interpretation of classical probability, we cannot arrive at a determination of a justifiable role for it. . . . The axiomatic reformulations remove some of the measurement ambiguities but do little to advance the problem of justification.

Logical Probability. Formal processing of empirical statements need not lead to empirically valid conclusions. . . . Carnap . . . attempted to justify logical probability as being valuable for decision-making, but good decision-making requires more than just coherence, if it even requires that. Hence the measurement of logical probability and the justification of an application of the theory are as yet unsolved; the former appears more likely to be settled than the latter.

Subjective Probability. Of all the theories we have considered, subjective probability holds the best position with respect to the values of probability conclusions, however arrived at. . . . Unfortunately, the measurement problem in subjective probability is sizable and conceivably insurmountable. . . . The conflict between human capabilities and the norms of subjective probability often makes the measurement of subjective probability very difficult. (pp. 238–239)

Trying to explain the outward success of the probabilistic approach in physics, Fine says that these results are

1) irrelevant to inference and decision-making,
2) assured by unstated methodological practices of censoring data and selectively applying arguments,
3) a result of extraordinary good fortune. (p. 245–246)

And now a few words of Tutubalin's criticism. His constantly repeated criticism (see, e.g., Tutubalin, 1972) is not so depressing as that by Fine. However, reviewing the book by Fine, he writes:

> By now rather a spicy situation has formed, when many popular textbooks, following the tradition which dates back to *Analytical Theory of Probabilities* by Laplace, greatly exaggerate the significance and sphere of application of probability theory. . . . far from all such alluring achievements, its declarations are fraught with significant negative scientific consequences, simply because the authors of textbooks do not guide themselves by their declarations in concrete actions; . . . But if everything usually written in textbooks is first earnestly accepted and then critically analyzed, the results will still be discouraging.

As a matter of fact we are facing an odd situation. On the one hand, a broad and, it seems, fruitful development of statistical methods is going on—probabilistic thinking and its effect upon scientists' views are being discussed. On the other hand, we hear disappointing warnings from some mathematicians. How can this be accounted for?

I believe that theory—or, better, theories—of probabilities promoted the creation of a probabilistic language. The proper mathematical constituent of these theories is their mathematical *structure*, *grammars* of dialects of the probabilistic language. And for this reason the problem itself of logical foundations on which to base the legitimacy of applying a specific probability theory seems quite meaningless. It is more fruitful to speak of using the probabilistic language to describe phenomena of the external world, this language being significantly softer than the traditional one, based on causal relations. Generally speaking, language can be convenient or inconvenient to describe something. The legitimacy of what we say in any language is given not by the structure (grammar) of the language but by the way we support our statements. Grammar serves only to make the phrases correct, i.e., corresponding to the rules of inference. Language is certain to influence the peculiarities of our argumentation. On some occasions it requires rigid causal relations, and on others it allows us to confine ourselves to vague but, somehow, normed judgments. A conversation without *any* rules seems to lack sense altogether.

We will start to develop our idea of a probabilistic language from the history of familiar deterministic conceptions of the world and the role this system ascribed to chance.

History of Determinism

Determinism as a concept has two meanings. Broadly, it is an unconditional belief in the power and omnipotence of formal logic as an instrument for cognition and description of the external world. In the narrow sense, it is the belief that all phenomena and events of the world obey causal laws. Furthermore, it implies confidence in the possibility of discovering, at least in principle, those laws to which world cognition is reduced.

The causal interpretation of the phenomena of the external world seems to be characteristic of the earliest forms of human thinking. At least, primitive tribes observable at present, with their rather alien forms of pre-logical thinking, lack the notion of chance altogether. To them, everything is mutually interrelated and predestined; all phenomena are perceived as signs or symbols of something. Levy-Brühl (1931), a well-known ethnologist of the recent past, describes this system of ideas:

> . . . For spirits so disposed, there is no chance; they do not overlook what we call the fortuitous. But as for a true accident, a sorrow, small or great, is never insignificant, it is for them always a revelation, a symbol, and has its reason in an invisible power which thus manifests itself. Far from being due to chance, it itself reveals its cause.

If these observations are extrapolated into the past, it allows us to believe that humanity at its early stages perceived and described the world in the language of irrational or even mystic causal notions.

The well-known ancient Indian teaching of Karma, as a large-scale system rigidly and meticulously determining man's fate through his actions in previous lives, may be regarded as a concise remnant of a once vague conception of the universal *causal* basis of the world.

Unfortunately, we do not possess the data which would allow us to trace the whole complicated progress of human thought which transformed ambiguous mystic ideas of causal relations into the logical structures of European thought. It is important to note that Aristotle classified and codified logical forms of expressing thought.[4] Later, in the Middle Ages, almost all intellectual life was devoted to the attempt to comprehend the role of logic in the universe and in human thinking. The

---

[4] Here is how Aristotle estimated the cognitive role of maintaining causal relations (quoted from Akhmanov, 1960, p. 159): "We believe we know every thing in a simple way when we believe we know the cause of its existence, and know not only its being the cause, but also that it cannot be otherwise."

Cause-Effect Links

medieval scholastics are responsible for the development and strengthening of rigid determinism.[5] They popularized (as well as vulgarized) Aristotle and made the study of logic universal. A very significant role was

[5] It may be supposed that determinism is deeply rooted in the widely known Cabala of the Middle Ages; this peculiar algebra of belief was built as a calculus over 22 letters–symbols, each of which corresponded to names and numbers. But who can trace the way this secret teaching influenced European thinking? However, many important analogies do arise, e.g., with the universal symbolism of Leibniz. Interest in the Cabala in the Middle Ages seems to have stimulated the construction of logical automata, of which so many beautiful tales are told.

played by Thomas Aquinas, the founder of Thomism, which was one of the two dominant trends of scholasticism. Here is one description of his outlook (Dragunov, 1970):

> Thomas defined the truth as "adequate correspondence (adequatio) between mind and thing . . ." (per se notum) and an unconditional principle of thinking and being, as well as a criterion for truly rational cognition. (p. 381)

The reader is here facing a very strong, though fairly naive, postulate of consistency both for thinking and for the processes of the world. Such axiomatics makes cognition and description of the world much easier, especially if the first postulate is accepted and interpreted as a primitive version of the reflection theory of truth accepted by Marxism.

The following words by William of Occam, a scholastic of the early fourteenth century who represented late nominalism, are also of interest (Styazhkin, 1967):

> Logic, rhetoric and grammar are not speculative subjects but genuinely cognitive guides since they really govern the mind in its activity. (p. 143)

The role of Thomas Aquinas in the evolution of European thinking was fixed by Pope Leo XIII in his encyclical "Acterni partis" in 1879, where Aquinas's philosophic-theological system was acknowledged as "the only true philosophy of catholicism" (Subbotin, 1972). It is probable that this papal intervention also contributed to the deification of a rigid determinism.[6]

An important role in understanding causality was played by Kant. According to him, space, time, and causality are prior forms of pure intellect, inherent categories which make our experience possible. Was this not the insight that causal and space–time arrangements of the observed phenomena are precisely a result of our language?

I next quote a very sharp statement concerning space, time, and causality made by A. D. Aleksandrov, a well-known contemporary physicist, in a paper devoted to philosophical comprehension of relativity theory (Aleksandrov, 1973):

---

[6] It is noteworthy that unofficial European religious–philosophical thinking, including the esoteric schools, is based upon the belief in rigid determinism. For example, in the beginning of the twentieth century there appeared the exposition of ostensibly ancient Egyptian teaching (previously concealed) where determinism was proclaimed, among other principles (Stranden, 1914, pp. 72–73): "Every cause has its effects; every effect has its cause; everything goes according to the law; chance is only a name by which we call the laws unknown to us; there are many aspects of causality, but nothing escapes the law." It is hardly wise to try to discover the true age of these statements. But, judging by their context, we may assume them to be a modernized and Europeanized exposition of some ancient oral concepts. I have quoted these words, unusual for a scientific paper, only to show to what extent determinism had penetrated European thinking.

The space-time structure of the world is nothing but its causal structure, but taken in a correspondingly abstract form. This abstraction
consists in omitting all the properties of phenomena and their causal
relations, except those indicating that phenomena are made of events
and their mutual effects are accumulated from the influence of some
events upon others.

This quotation, certainly, is not a direct return to the Kantian notion
of the inherent nature of space and time, since the notion of inherence
here has a different — modern — sense: inherent is the capacity for profound abstraction. It is easy to assume that this capacity developed in the
process of evolution and became genetically fixed, and in this sense (and
not in some metaphysical–idealistic one) it became inherent. But the
most important feature here is that if, in dealing with notions of space,
time, and causality, we have to acknowledge the existence of certain universal structures of our perception, which arise as a result of abstraction,
we thus acknowledge their linguistic nature. Without these structures of
perception we could not have discussed our observations of the external
events. This immediately gives rise to the question: Are we dealing with
the only possible system of structuring or are other linguistic categories
also possible?

Now let us consider the role of determinism in the development of
science. It began with classical mechanics, the simplest theory based
upon determinism (at least so it seems at first glance). From school years
we are brought up with the idea that by applying the laws of classical
mechanics it is possible to predict the future of a material system, its initial data being known. Later, we reach the conclusion that classical
mechanics is the best example of our knowledge.

As a matter of fact, everything we obtain by means of classical
mechanics is nothing more than an approximate description. Strictly
speaking, initial data are never known with certainty; the only information which may be evident is their distribution. Further, during motion
unpredicted random forces can influence a system; at least, it is not likely
that a given system will remain isolated during the period in which we are
going to make predictions. Even such an accurate branch of science as
celestial mechanics needs corrections from time to time. All this is well
analyzed in a highly readable book by Blokhintsev (1966). I shall only
remark that classical mechanics has holes in a purely theoretical sense.
Bohr (1955) drew attention to the fact that the law of inertia violates the
principle of causality: a uniformly moving body in a vacuum keeps moving without any cause.

However, when the laws of classical mechanics were applied, the approximation often proved so accurate that it struck scientists as a
miracle. This high degree of accuracy was accounted for by the fact that

people succeeded in selecting phenomena invariant in relation to the system in which they take place. Wigner (1960) believes that Galileo must have been puzzled by the fact that stones fall from a tower in a manner independent of their size, the weather, or who throws them. Classical physics, too, studied the phenomena invariant in relation to the changing states of the system. One of the favorite questions of today asks whether chemistry or even biology is reducible to physics. But it turns out that not even all physical problems can be reduced to a description in the framework of traditional physical concepts.

This is true, first of all, of problems of technical physics, where the object of research is the system itself. One of my favorite examples of such systems is sprectrochemical analysis. Here we are dealing with the system in which many well-known physical phenomena act simultaneously: hydrodynamic flow by discharge, explosion evaporation, equilibrium evaporation, selective oxidation on the electrodes, diffusion in a solid (influenced by the solid's structure), emergence of a gas cloud, diffusion within it, excitation of atoms and radiation — and on top of all this, inaccurate sharpening and installing of electrodes and unstable parameters of the excitation generator. In this system, invariants — dominant phenomena — cannot be selected. Nothing can be described in familiar terms of physics, though it is always possible to think of an experiment in which nearly every one of the enumerated phenomena could be considered almost as an invariant. However, the problem is formulated so as to enable the study of the whole system.

True, still earlier physicists had to face the impossibility of selecting dynamic invariants in constructing the kinetic theory of gases; in order to connect molecular processes with the macroscopic state of a system, they had to introduce a probabilistic description.

Later, in quantum mechanics, the change of scales made the concept of the precise particle location in space impossible, whence comes the impossibility of the familiar notions of phenomena arranged in time and space.

But all this is known only too well. I remind the reader of the evolution of thinking in physics only in order to answer the question analogous to that asked by Fine in discussing the legitimacy of probabilistic conceptions for description of the external world. The question can be stated as follows: Have we sufficient logical grounds to describe the external world with deterministic concepts? These grounds, if any, are more of a historical–psychological nature than a logical one. Human prehistory prepared people for causal interpretation of phenomena. Medieval scholastics strengthened and deepened belief in "determinism" in its broad meaning. Progress in classical mechanics fixed the belief into the causal picture of the world for a long time. Later scientific development, especially in physics, continued to contribute to this belief, but eventually it began to shake it loose.

If we wish to justify rigid determinism in its broad sense from a strictly logical position, we have to recognize as axioms Thomas Aquinas's statements that thinking is consistent, the world is consistent, and consistency is the criterion of the truth. Hilbert's program, directed at proving the absolute consistency of mathematical structures, and the program of the neopositivists in the form in which it was set forth by Carnap — are these not a distant echo of the postulates of Thomas Aquinas? After both these programs failed, and especially after the appearance of Gödel's theorem on undecidability and the progress of quantum mechanics, what can we say in favor of the absolute belief in determinism? Although I am quite aware of the fact that certain outstanding scholars of the recent past, say, Einstein, were determinists (his discussion with Bohr on the subject is well known), it seems more pertinent to speak here of the paradigm of the epoch rather than of clearly formulated logical foundations.

History of the Teaching of Chance

I have already pointed out that the concept of chance was quite foreign to the psychology of primitive people. It is impossible to trace in any detail the history of the emergence and formulation of such concepts as probability and chance. Only scanty information is available.

The intellectually rich society of Ancient India lacked the concept of probability in its modern meaning, though Indian thinkers understood only too well the universal changeability of the world and approached rather closely contemporary ideas of stochastic processes. This can be illustrated by the famous dialogue between Milinda and Nagasena (Oldenberg, 1881). This dialogue is a fragment of an historical document which records the account of the meetings between Menander (Milinda), a Greek prince who ruled on the territory of the Indus and in the valley of the Ganges in 125–195 B.C., and Nagasena, a Buddhist teacher.

"It is as if, sire, some person might light a lamp. Would it burn all night long?"

"Yes, revered sir, it might burn all night long."

"Is the flame of the first watch the same as the flame of the middle watch?"

"No, revered sir."

"Is the flame of the middle watch the same as the flame of the third watch?"

"No, revered sir."

"Is it then, sire, that the lamp in the first watch was one thing, the lamp in the middle watch another, and the lamp in the last watch still another?'

"O, no, revered sir, it was burning all through the night in dependence on itself."

"Even so, sire, a continuity of dhammas runs on, one uprises, another ceases; it runs on as though there were no before, no after; consequently neither the one (dhamma) nor another is reckoned as the last consciousness."

What can be said of the causal arrangement of phenomena in time and space if such an outlook is shared? Does this not resemble a modern description of a random process?

The well-known Indian statistician Mahalanobis tried to trace a certain analogy between modern statistical theory and the ideas of Jainist logic [the religion and philosophy of Jaina, which reached its full blossom in the times of Great Mahavira (589–527 B.C.), Buddha's contemporary]. Jainism contained a system of ideas called syādvada, close to the modern probabilistic concepts (Mahalanobis, 1954).

> The *syādvada* is set forth as follows: (1) May be, it is; (2) may be, it is not; (3) may be, it is and it is not; (4) may be, it is indescribable; (5) may be, it is and yet is indescribable; (6) may be, it is not and it is also indescribable; (7) may be, it is and it is not and it is also indescribable.
>
> . . . all things are related in one way or the other and . . . relations induce relational qualities in the relata, which accordingly become infinitely diversified at each moment and throughout their career. . . . Things are neither momentary nor uniform.
>
> A reality is that which not only originates, but is also liable to cease and at the same time is capable of persisting. Existence, cessation, and persistence are the fundamental characteristics of all that is real. . . . This concept of reality is the only one which can avoid the conclusion that the world of plurality is the world of experience, is an illusion. (p. 103)

One cannot help wondering, while reading these ancient fragments, at how far they had progressed as compared with primitive notions of simple space–time arrangement. And how far is the conception of syādvada from the limitations imposed upon thinking by the laws of formal logic. Still, all this coexisted with the universally accepted notion of karma — a system most rigidly arranged on a large time scale.

It would be very interesting to trace the pre-history of probabilistic notions in European thinking, but this is very difficult to do. The point is that neither the Hellenic epoch nor the Middle Ages gave birth to any coherent probabilistic conceptions. There occurred only separate and often contradictory statements about the role of chance, which could have many interpretations. Still, I shall try to show how elements of probabilistic judgments began to ooze through a general deterministic background. I base my argument principally upon the very interesting and thorough paper by Sheynin (1974).

It is difficult to say anything definite about the way the role of chance was estimated by Greek atomists. They were strict determinists. Democritus rejected chance quite obviously. Well known is the statement by Lysippus: "Nothing comes from nothing but everything comes from foundations and necessity." At the same time, the atomists were reproached by their contemporaries because they would ascribe everything to chance since this logically followed from their constructions [for more details, see Russell (1962, p. 85) and Sheynin (1974, p. 102)].

Chance is repeatedly mentioned by Aristotle. According to Sheynin, it was Aristotle who introduced the concept of chance and accident into classical philosophy, defining it as follows: accident "is something which may possibly either belong or not belong to any one and the selfsame thing . . ." (Sheynin, 1974, p. 98). Sheynin draws our attention to the fact that Aristotle's works also contain reasoning on the probable. A probability, says Aristotle,

> is a generally approved proposition: what men know to happen or not to happen, to be or not to be, for the most part thus and thus . . ., e.g. "the envious hate" . . . (Sheynin, 1974, p. 101)

Below are several more statements by Aristotle, which I cite as they are given by Sheynin (1974):

> As to chance (and change) they are "characteristic of the perishable things of the earth" . . . Some effects could be caused incidentally, i.e. by spontaneity and chance, chance is opposed to mind and reason and its cause "cannot be determined. The products of chance and fortune are opposed to what is, or comes to be, always or usually." (p. 98)

The general impression is that Aristotle, acknowledging the role of chance in life, attributed it to something which violates order and remains beyond one's scope of comprehension. He did not recognize the possibility of a science of chance, though he understood that various human activities are connected with it. Aristotle said that in navigation "not the cleverest are the most fortunate, but it is as in throwing dice" (p. 101). He described rhetoric as an art of persuasion based on probabilities. As Sheynin (1974) points out, he even introduced a rudimentary scale of subjective probabilities stating that "a likely impossibility is always preferable to an unconvincing possibility." (p. 101).

Now let us see what Thomas Aquinas's attitude toward chance was.

Here are several statements from his famous tractatus *Summa Theo-logica* [again cited from Sheynin's article (1974)]:

(1) The effects willed by God happen contingently . . . because God has prepared contingent causes for them.
(2) Casual and chance events are such as proceed from their causes in the minority of cases and are quite unknown. (p. 103)

. . . some causes are so ordered to their effects as to produce them not of necessity but in the majority of cases, and in the minority to fail in producing them . . . which is due to some hindering cause . . . (p. 103)

. . . if we consider the objects of science in their universal principles, then all science is of necessary things. But if we consider the things themselves, then some sciences are of necessary things, some of contingent things. (p. 104)

The conception of chance shared by Aquinas is, of course, hard to outline very clearly. Sheynin remarks that, according to Byrne (1968), there is something in common between Aquinas's conception of probability and modern logical probability theory as well as between his theory of contingency and the modern frequency theory. However, be that as it may, Aquinas's views on the nature of chance did not influence the progress of probabilistic concepts in modern times. The quotations cited above are interesting for us in that they illuminate the way an outstanding thinker of his times, an ardent believer in logic, tried to cope with chance.

Medieval scholastics also faced the necessity to comprehend another concept in the framework of logical notions: "free will." Logically speaking, the concept of free will is equivalent to that of chance. If in an experiment with tossing a coin many times, we assume that in each given fall, in accordance with the concept of chance, the coin lands unpredictably, this is logically equivalent to ascribing to it free will with certain well-known statistical limitations laid upon the set of tosses of the coin. Free behavior is as difficult to ground logically as the possibility of chance is. Not without reason is the problem of free will one of the "accursed" questions of philosophy.[7] Already Buridan, a French scholastic, believed the problem of free will to be logically unsolvable. His argu-

[7] Attempts to find logical grounds for the concept of freedom have been made in our times, too. This is illustrated by an elegant paper by Gill (1971), where freedom is considered in the framework of a calculus — in order to define freedom, three postulates are introduced, and two theorems are formulated. In constructing a calculus, chance, naturally, has to be excluded. This is formulated in the following manner: "If a given command is contingent, its contradictory opinion is necessary—it cannot be rejected without self-contradiction. The agent cannot control himself if he commands or permits an inconsistency. If a prohibition is contingent — not necessary—its contradictory permission is necessary" (p. 9). This is a brilliant sample of strictly formal reasoning in the style of logical positivism, showing that chance makes a system internally inconsistent.

ments seem to have influenced the later development of the problem. In any case, a well-known paradox is that of the *pons asinorum* (falsely ascribed to Buridan) about the donkey starving to death between two bales of hay as a result of the absence of logical grounds for decision making. The possibility of a random choice is here excluded as illogical.

Among modern philosophers, many outstanding thinkers either did not acknowledge the role of chance, or if they did, they connected it with numerous unknown causes. For example, the Dutch philosopher B. Spinoza (1632–1677) wrote:

> *Prop. XXIX.* Nothing in the universe is contingent, but all things are conditioned to exist and operate in a particular manner by the necessity of the divine nature.
> *Proof.* Whatsoever is, is in God, for he exists necessarily, and not contingently. Further, the modes of the divine nature follow therefrom necessarily, and not contingently . . . (Spinoza, 1955)

The British philosopher Thomas Hobbes wrote:

> (1) . . . generally all contingents have their necessary causes . . . but are called contingent in respect of other events upon which they do not depend . . .
> (2) . . . by contingent, men . . . mean . . . that which hath not for cause anything that we perceive . . . (cited from Sheynin, 1974)

The French philosopher C. A. Helvetius (1715–1771), in his famous tractatus *On Mind*, wrote:

> . . . chance; that is, an infinite number of events, with respect to which our ignorance will not permit us to perceive their causes, and the chain that connects them together. Now, this chance has a greater share in our education than is imagined. It is this that places certain objects before us and, in consequence of this, occasions more happy ideas, and sometimes leads to the greatest discoveries. . . . If chance be generally acknowledged to be the author of most discoveries in almost all the arts, and if in speculative sciences its power be less sensibly perceived, it is not perhaps less real . . . (Helvetius, 1809, p. 221)

According to the French philosopher P. H. T. Holbach (1723–1789):

> . . . Chance, a word devoid of sense , which we always oppose to intelligence without coupling it with a clear idea. In fact, we attribute to chance all those effects concerning which we see no link with their causes. Thus, we use the word chance to cover our ignorance of the natural causes which produce the effects that we see. These act by means that we have no idea of or they act in a manner in which we do not see any order or system; followed by actions similar to our own. Whenever we see, or believe we see order, we attribute this order to

an intelligence, a quality derived from ourselves and from our fashion of acting and being affected. (Holbach, 1770)

We read in the works of the British philosopher David Hume (1711–1776):

> . . . chance is nothing real in itself, and, properly speaking, is merely the negation of a cause. . . . it produces a total indifference in the mind . . . the chances present all these sides of the die as equal, and make us consider every one of them, one after another, as alike probable and possible. . . . the chance or indifference lies only in our judgement on account of our imperfect knowledge, not in the things themselves, which are in every case equally necessary. (Hume, 1964)

Immanuel Kant wrote:

> In a body these absurdities were taken to such an extreme that they ascribed the origin of all living creation precisely to this blind concourse and actually derived reason from unreason. In my own concept, on the other hand, I find matter/substance bound to certain, distinct, necessary laws. . . . there exists a System of all Systems, a limitless understanding, and an independent Wisdom from which Nature also derives her origins according to her possibilities in the entire sum of determinations. (Kant, 1912)

Kant's *Critique of Pure Reason* contains a statement already acknowledging chance, at least in its individual manifestation: ". . . the individual accident (chance) is nevertheless entirely subordinated to a principle (rule)." This is, if you like, a concession to chance made within the boundaries of determinism. The German physician, naturalist, and philosopher L. Büchner (1824–1899) wrote:

> What we call chance is exclusively founded upon the tangle of circumstances, whose inner relations and final causes we cannot discover. (Büchner, 1891)

Hegel made a much more profound attempt to comprehend chance. In his *Science of Logic* we find the following statements:

> This union of Possibility and Actuality is Contingency. [The Contingent] has no foundation. The Contingent is indeed Reality as only that which is possible . . . this has a foundation.
>
> The contingent therefore, in consequence, because it is accidental has no ground, but even so it has a ground just because it is accidental. Here the union of necessity and contingency is itself present; this union is called Absolute Reality. (Hegel, 1971)

The words of Hegel, as usual, cannot be comprehended completely. But it is important to note that they lack naive negligence of chance and its reduction to uncomprehended or undiscovered causes. Engels wrote:

. . . where on the surface accident holds sway, there actually it is always governed by inner hidden laws and it is only a matter of discovering these laws. (Engels, 1973, p. 48)

There was a time when Soviet philosophical literature displayed an acutely hostile attitude toward chance as a philosophical category. For example, *Concise Dictionary of Philosophy* edited in 1955 contained the following opinion (Rosental and Yudin, 1955):

Cognition may be considered scientific only so far as it acknowledges the natural and social phenomena in their necessity. Cognition cannot be based on randomness. Behind randomness science always strives to discover regularity and necessity (pp. 325–326).

Later, such extreme judgments were recognized as erroneous. Now the following statement, given in the *Philosophical Encyclopaedia* published in 1970, seems to be considered correct:

Science by no means stops at randomness but strives to understand regularity and necessity. But recognizing the objectivity of randomness, we have to recognize the necessity to study it. Random phenomena and processes are a special object of several modern sciences, including physics, biology, sociology, etc. Such branches of modern mathematics as theory of probability, theory of random functions, theory of stochastic processes, are completely devoted to studying quantitative characteristics of chance (Yakhot, 1970, p. 34).

This text is already a significant step forward. Science gets the right to study chance, though it is said that it does not stop there, but strives to understand regularity. It only remains unclear how to pass from quantitative parameters to understanding necessity.

The naive belief that chance emerges in our consciousness as a consequence of ignorance passed from philosophy to the natural sciences, where it was shared by Galileo, Kepler, Huygens, Bernoulli, Lambert, and even Laplace. Here is what Kepler said of chance:

But what is Chance? Nothing but an idol, and the most detestable of idols—nothing but contempt of the sovereign and all-powerful God as well as the very perfect world that came from his hands. (cited from Sheynin, 1974, p. 127)

And here are the words of Laplace, one of the creators of probability theory:

Chance has no reality in-itself; it is nothing but the proper terms for designating our ignorance of the manner by which the different parts of a phenomenon coordinate among themselves and the rest of Nature. (cited from Sheynin, 1974, p. 132)

I would also like to mention the thoughts of two outstanding scholars of comparatively modern times on the subject of chance. Darwin in his *Origin of Species* wrote:

> I have hitherto sometimes spoken as if the variations . . . had been due to chance. This, of course, is a wholly incorrect expression, but it serves to acknowledge plainly our ignorance of the cause of each particular variation. (cited from Sheynin, 1974, p. 115)

Henri Poincaré, one of the most prominent mathematicians of the recent past and one of the first scholars interested in the philosophical foundations of science, also believed [as pointed out by Sheynin (1974)] that chance has an influence when, under the conditions of unstable equilibrium, very weak causes produce a very strong effect (see Poincaré, 1952).

The concept of chance was initially introduced into science by physicists, at the end of the twentieth century. They seemed to feel quite unmoved by the problem of the philosophical comprehension of chance. They had to explain and describe the world, and this description did not fit the limits of deterministic conceptions. Certain phenomena could only be well described in probabilistic language. The landmarks of this process are well known: creation of kinetic theory of matter by Maxwell and Boltzmann; the latter's statement that our world is but a result of a huge fluctuation; introduction of the notion of an ensemble by Gibbs and the canonical distribution discovered by him (this led not only to the creation of statistical physics but to something more—to forming a new outlook in physics); the study of Brownian motion, which gave impetus to developing the theory of random functions; and, at last, the progress of quantum mechanics. But they were not worried about the philosophical problem or logical foundations of the legitimacy of this approach. The world of observed phenomena was well described, and this was a sufficient foundation. The sorrowful ponderings of the philosophers of the past about chance were merely forgotten. Here are the thoughts of the well-known physicist Max Born on the relation of randomness and determinism:

> We have seen how classical physics struggled in vain to reconcile growing quantitative observations with preconceived ideas on causality, derived from everyday experience but raised to the level of metaphysical postulates, and how it fought a losing battle against the intrusion of chance. Today the order of ideas has been reversed: chance has become the primary notion, mechanics an expression of its quantitative laws, and the overwhelming evidence of causality with all its attributes in the realm of ordinary experience is satisfactorily explained by the statistical laws of large numbers. (Born, 1949, p. 120–121)
>
> . . . I think chance is a more fundamental conception than causality;

for whether in a concrete case a cause-effect relation holds or not can only be judged by applying the laws of chance to the observations (Born 1949, p. 47)

However, this is only a panegyric to chance; in no way is it a logical analysis of what chance is.

I shall not here dwell on the development of probabilistic concepts in mathematics. The early period—the eighteenth century and the beginning of the nineteenth century—is thoroughly illuminated in the papers by Sheynin (1971*a*, *b*, 1972*a*, *b*, 1973*a*, *b*, *c*). The later period is well known to everybody who is interested in probabilistic concepts. I shall only make one brief remark. When probabilistic methods in mathematics began to develop, they proceeded not from some general concepts of the insufficiency of deterministic methods to describe the phenomena of the external World (analogous, say, to the philosophy of Jainism) but from the attempt to describe and comprehend two quite particular phenomena: on the one hand games of chance and on the other hand elaboration of the theory of errors which resulted from the introduction of degree measurements in the instrumental astronomy in the seventeenth and eighteenth centuries (for details, see the papers by Sheynin).

Mathematical statistics in its modern form was created only at the end of the nineteenth and the beginning of the twentieth century, after the publication of papers by F. Galton, K. Pearson, and R. Fisher. Then probabilistic methods of research began to penetrate into various fields of knowledge.

Formation of a Probabilistic Paradigm

The theory of probability or, better, theories of probabilities of the present create something more than a theory for describing mass, repeated phenomena: they generate a new paradigm that allows one to describe the observed world in a weaker language than that of the rigid deterministic ideas traditionally accepted in science.

Language of probabilistic concepts.   I shall try to elucidate this idea in detail. We say that a random value is given if its distribution function is given. That means that we quite consciously abandon the causal interpretation of the observed phenomena. We are satisfied with a purely *behavioral* description of phenomena. A distribution function is a description of random value behavior, without any appeal to what has caused this type of behavior. At last, we acquire the right to describe a phenomenon simply as it is. Moreover, the description is given in some blurred, uncertain way: the probability that a continuous random value in its realization (say, as a result of measuring) will occupy some fixed point equals zero. We can speak only of the probability that some random value will fall within an interval of values.

Does not this imply quite a novel view of the world or, at least, the possibility of a description radically different from the traditional deterministic one?

Let us examine the well-known illustration with tossing a coin. Remaining in the probabilistic position, we assume that in each separate tossing a coin may fall as it likes; i.e., as I have already said, we ascribe to the coin free will, though we also lay statistical limitations upon the result of large numbers of tests. This is a much weaker description of a phenomenon than an attempt to predetermine, proceeding from the laws of mechanics, in what way the coin will fall. At first sight, it seems that the chain of causal phenomena leading to a concrete result in a concrete act of tossing the coin may be traced in the main. But if we try to do this, we shall immediately have to introduce into consideration an incredibly large, perhaps infinitely large, number of facts and circumstances, and our chain of causal links will have to be extended to include space phenomena rooted in some immensely remote past that is unknown to us.[8] It is noteworthy that tossing a coin is almost the same as the experiment with which Galileo started the progress in mechanics. However, in one formulation of the problem, the experiment with throwing proves invariant to the surrounding phenomena, whereas in another formulation this is not so.

All that was said above pertains not only to tossing a coin or dice. It also pertains to the behavior of error in any experiment as well as to the behavior of any sufficiently complicated system. As mentioned above, Darwin thought an attempt to explain variation in biology by chance should not be taken seriously. However, at present we have every reason to believe that the origin of species cannot be regarded as a result of a rigidly given program. Mutations have to be connected with chance. This follows both from biological considerations (Monod, 1972) and from logical ones (Nalimov and Mul'chenko, 1970; see also Chapter 7 of this book). From Gödel's proof of undecidability, it clearly follows that any sufficiently rich logical system is incomplete, and extended, but finite,

---

[8] It is of interest to quote here statements by Max Born concerning the difficulty of understanding the idea of a "causal chain" (Born, 1949, p. 129): "One often finds the idea of a 'causal chain' A_1, A_2, . . . where B depends directly on A_1, A_1 on A_2, etc., so that B depends indirectly on any of the A_n. As the series may never end, where is a 'first cause' to be found? — the number of causes may be, and will be in general, infinite. But there seems to be not the slightest reason to assume only one such chain, or even a number of chains; for the causes may be interlocked in a complicated way, and a 'network' of causes (even in a multidimensional space) seems to be a more appropriate picture. Yet why should it be enumerable at all? The 'set of all causes' of an event seem to me a notion just as dangerous as the notions which lead to logical paradoxes of the type discovered by Russell. It is a metaphysical idea which has produced much futile controversy."

expansion of its axioms does not make it complete. In the language of such a system, true statements may be formulated which do not immediately follow from it, as well as false statements which will not be refuted. A deterministic description of the world as a whole, or even of a large subsystem such as the biosphere, must remain impossible. Yet, a consistent description of the world by appeal to chance seems intuitively possible.

The impossibility of accurately locating a particle, revealed in quantum mechanics, also mandates a "blurred" description of observed phenomena by means of probability waves. It is as a consequence of this weakened type of description that the causal nature of the system's progress can be preserved. In the words of Born (1949):

> [in quantum mechanics] we have the paradoxical situation that observable events obey laws of chance, but that the probability for these events itself spreads according to laws which are in all essential features causal laws. (p. 103)

The introduction of probability waves in quantum mechanics is, if you like, just the weakening of the rigid causal concepts of classical physics. The development of a wave is predictable during the observation, but prediction itself is of a non-deterministic nature to which we are accustomed in everyday life. The logic of reasoning is such that the causal progress of events is not completed. It breaks somewhere and is replaced by a probabilistic description of behavior.

An algorithmic definition of randomness as the complexity of a message can also be interpreted as a behavioral description. If we deal with a sequence of numbers consisting of zeros and ones, then, roughly speaking, complexity will be characterized by the minimal number of binary digits necessary to replace the sequence in transmitting it through a communication channel. According to A. N. Kolmogorov, those elements of a large finite aggregate of symbols are called random which have the greatest complexity. The concept of randomness emerges here from observing the behavior of a symbolic sequence. If it is impossible to discover an algorithm for generating numbers which would be simpler than the sequence, then the whole sequence must be transmitted through the communication channels. Such a sequence is naturally called random.

Fine (1973) tries to contrast determinism to chance in the following manner:

> We can distinguish between deterministic and chance phenomena capable of generating an indefinitely long sequence of discrete-valued outcomes on the grounds that deterministic phenomena yield outcomes of bounded complexity, whereas chance phenomena yield

outcomes for which the complexity of increasing longer outcomes di-
verges. Probabilistic phenomena might then be characterized as the
subset of chance phenomena for which the various outcomes have apparently convergent relative-frequencies. (p. 153)

Any algorithmic definition of a random sequence is clearly linguistic.
Roughly speaking, we call random what we cannot describe briefly. And
this is where language relativism immediately shows up. Imagine that we
are dealing with numbers π and e. It is clear that there is no necessity to
transmit through a communication channel all the figures giving the approximate value of these numbers: it will suffice to give the algorithm of
their calculation. In this sense symbolic sequences approximately giving π
and e are not random. At the same time it is known that these sequences
of numbers are sometimes used as random ones in problems of simulation by the Monte-Carlo method. Indeed, statistical criteria we have at
our disposal do not allow us to differentiate these sequences from those
given by a meter registering radioactive decay. Now imagine that a symbolic sequence of π is recorded with the first symbols omitted. Who will
guess that the sequence is not random? (True, this is not the only
trouble.) The algorithmic approach is fraught with difficulties resulting
from a particular choice of calculation programs. The conception, as a
whole, is far from being complete.

Besides Kolmogorov's definition there are also definitions of randomness for infinite sequences, given by Donald W. Loveland, P.
Martin-Lof, and G. J. Chaitin. Several definitions of probability based
on the evaluation of complexity are proposed by R. J. Solomonoff. I
have noted Kolmogorov's statements on this point. A more detailed
discussion of the difficulties connected with the elaboration of
algorithmic randomness is presented by Fine (1973), whose book also
contains a substantial bibliography on the subject.

Axiomatics of the theory of probability as grammar.   If probability
theory in its applications is regarded as a language, its structure, given by
the axiomatics, will be just the grammar of this language (Nalimov,
1974*a*; see also Chapter 3 of this book). By this approach we immediately
avoid all Fine's (1973) lamentations that, from the foundations of probability theory, nothing follows concerning the possibility of its application. Any grammar, according to the meaning of the word, is aimed only
at constructing grammatical and comprehensible—meaningful and consistent, or at least roughly consistent—phrases. But from grammar
nothing ever follows concerning a language's applicability.

It may seem that the axiomatics of probability theory [we shall consider here only generally acknowledged axiomatics (Kolmogorov, 1956)]
is, indeed, used as grammar; i.e., one has to fall back upon it while constructing comprehensible phrases. I would like to illustrate this statement

by some examples. When a probabilistic statement is made, it is first of all necessary to be aware of the space of elementary events on which the probabilities are given. Otherwise, we can get absurd results such as probability greater than unity.[9]

The concept of σ-algebra gives a clear idea of the set of elementary events under consideration. One of the requirements here is: if A belongs to the set of events, then $\overline{A}$ (i.e., not A) also belongs to it, which is to say that a grammatically correct system of statements is built so that all possible logical operations remain within σ-algebra.

This is important if one is to understand texts containing probabilistic judgments.

Axioms of norming and non-negativeness are of great importance for understanding probabilistic statements. If we come across a statement containing negative probability, it will simply remain incomprehensible. And if we try to record undetermined behavior of a phenomenon not in probabilistic notions but in some unnormed weight functions [as Zadeh (1971) does in his theory of fuzzy sets], the statements based on the record, though they will be understood, will have quite another meaning than the statements made in the probabilistic language. Here is an illustration. Assume that somebody, proceeding from certain non-probabilistic considerations, wants to write a formula analogous to the Bayesian one (in the system of notions of subjective probabilities), but for unnormed weight functions

$$p(\mu|y) = kp(\mu)p(y|\mu)$$

If the recorded functions are presented merely as unnormed weight, it will be natural if the coefficient k is to equal 1. Now assume that functions $p(\mu)$ and $p(y|\mu)$ are such that one of them reaches its maximum in one part of the abscissa and the other, in another part, the maximum of one function corresponding to a gently sloping curve of the other one, with ordinate values close to zero. The product of the two functions will yield a bimodal function $p(\mu|y)$ with small weight values for the peaks. We have to acknowledge that the character of the functions $p(\mu)$ and $p(y|\mu)$ makes us give rather an unaccustomed interpretation. It would look as if we are dealing here with a case of "twilight" consciousness when a person cannot clearly enough formulate his ideas. At the same time, if we share probabilistic views, dealing with the same functions $p(\mu)$ and $p(y|\mu)$ we shall obtain a bimodal function $p(\mu|y)$ normalized to 1, which will be familiarly interpreted (in terms of subjective probabilities)

---

[9] This idea can be illustrated by von Mises's paradox, which was presented earlier (see p. 23). Despite the obvious absurdity of the reasoning in this paradox, the old probability theory lacked anything which would forbid it.

in the following way: in the resulting judgment generated by mixing certain prior information with that received in the given experiment, we have two meanings, and both of them may have approximately equal probabilities. Such conclusions help to explain some seemingly unexpected events of real life.[10]

In criticizing Kolmogorov's axiomatics, Fine pays attention to the fact that two fundamental concepts of probability theory—independence of random values and conditional probability—remain irrelevant to axiomatic constructions: they are given by separate definitions, and Fine believes axiomatics to be incomplete in this respect. But if the structure of probability theory is regarded as the grammar of a language, this remark by Fine, interesting in itself, does not have any essential significance.

Also, if the axiomatics of probability theory is viewed as grammar, the question of its consistency is not of great importance either, and I shall not dwell upon it here[11] [on the growing tolerance to the problem of consistency in mathematics see, for example, Gnedenko (1969)].

In concluding this analysis of the axiomatics of probability theory, I have to acknowledge that not all of its rich content explicated in theorems is used as grammatical structures. Many fairly important theorems of probability theory, e.g., the law of repeated logarithm,[12] have no obvious grammatical interpretation. Mathematical structures, in their practical application, give the language grammar but are not reduced to it.

Physical interpretation of the concept "probability."   If the probability theory is considered from a linguistic point of view, then Fine's (1973) complaints, supported by Tutubalin (1972), that from axiomatic struc-

---

[10] For example, the Bayesian theorem helps to explain the nature of anecdotes in our everyday verbal behavior. Assume that $p(\mu)$ is a prior distribution function of the sense content of a highly polymorphous word. An anecdote may be constructed so that the given word, combined with others, generates in the listener's mind function $p(y|\mu)$ with the maximum in another part of the abscissa, in which the prior distribution function is gently sloping close to the abscissa. As a result, the posterior distribution function will prove bimodal. The anecdotal character of the situation will derive from the fact that the word may have two equally common but essentially different meanings—hence two meanings of the phrase. For more details, see my book *In the Labyrinths of Language: A Mathematician's Journey* (Nalimov, 1981).

[11] It is of interest to observe how the problem is treated in books on probability theory. Tutubalin (1972) formulates it but avoids its detailed discussion, referring only to the fact that the notion of a set, as it is used in constructing axiomatics, leads to paradoxes which cannot be overcome at present in a sufficiently satisfying way. Gnedenko (1969) gives the following argument for the consistency of Kolmogorov's axioms: "Kolmogorov's system of axioms is *consistent* since there exist real objects which these axioms satisfy" (p. 50). Such a basis of consistency, broadly accepted in the pre-Hilbertian period, presupposes acknowledging the above-mentioned postulate by Thomas Aquinas of the World's "consistency."

[12] A remarkable theorem by the Russian mathematician A. Ya. Khinchin specifying the Law of Large Numbers, well known in probability theory. This theorem has brought about a number of serious studies.

ture there does not follow an interpretation of the physical sense of probability, remain incomprehensible.

It is natural to believe — and this is generally accepted at present — that logical grammar deals with symbol systems independently of how they are interpreted in terms of the external word. Interpretation appears later, when language is used to formulate concrete statements. And this interpretation may be polymorphous and fuzzy. Kolmogorov (1956), after his shattering criticism of the conception of von Mises, still gives a frequency definition of probability, though, of course, without transition to the limit. He writes that it suffices to speak of probability as a number around which frequency is grouped under definitely formulated conditions, so that this tendency to grouping is manifested more clearly and accurately with the growing number of tests (up to a reasonable limit).

Such definitions of probability entered the textbooks, too. In Tutubalin's book (1972) we read, "The number around which the frequency of event A fluctuates, is called the *probability of event A* and is designated by $P\{A\}$" (p. 6).

We feel a desire to ask: Should this interpretation be considered as the only possible one? It is hard to believe that physicists who study quantum mechanics will agree to this.[13] It is altogether incomprehensible why we should exclude a consideration of probability as a measure of uncertainty in our judgments. If the concept of subjective probability is introduced (as it is by L. J. Savage, Bruno de Finetti, and other representatives of this trend), it proves possible to apply to it all the usual rules of probability calculus.

The requirement of statistical stability.   The frequency interpretation of probability immediately gives rise to the problem of stability, very acutely introduced by Richard von Mises. This problem is, if you like, a stumbling-block in discussing all the questions related to the applicability of probabilistic notions for describing external phenomena. On this point, Tutubalin (1972) writes:

> According to modern views, the area of application of probability methods is limited to phenomena characterized by their statistical stability. However, testing statistical stability is difficult and always in-

---

[13] Here is how the meaning of the wave function is treated by Blokhintsev (1966): ". . . the wave function is not a value determining the statistics of a special measurement; it is a value determining the statistics of a quantum ensemble, i.e., the statistics of any measurements compatible with the nature of microsystem μ and macroscopic situation M which dictates the conditions of movement for the microsystem μ." In his latest book, Blokhintsev (1978) proposes to denote by the term "probability" the measure of the potential possibility of an event's occurring. The American philosopher Abel has collected physicists' statements on the concept "wave function"; in a slightly contracted form it is given in my earlier books (Nalimov, 1974b, 1981).

complete; besides, it often leads to negative conclusions. As a result, in some branches of knowledge, e.g., in geology, it has become a norm not to test statistical stability, which often leads to serious blunders. (p. 144)

In the book by Fine (1973), we find a sad remark that the stability of frequencies, upon which the application of probability theory must be based in the problems of forecasting, in no way follows from Kolmogorov's axiomatics. In some textbooks on probability theory, stability of frequencies is ascribed almost the status of a law of nature. In the book by Ventsel (1962), we read:

. . . the property of "stability of frequencies," many times tested experimentally and supported by all the experience of human practical activities, is one of the most universal regularities observed in random phenomena. (p. 29)

In the book by Gnedenko (1969), we read,

Permanent observations over appearance or nonappearance of event A in a large number of repeated tests under the invariable complex of conditions show that for a broad circle of phenomena the number of appearances or nonappearances of event A obeys stable regularities. (p. 41)

I consider all these judgments on the stability of frequency to be a result of misunderstanding, to a certain degree. The concept of frequency stability is nothing more than a logical construction. Without this statement, it is impossible to give a limit-frequency interpretation to the notion of probability. Mathematically, the statement of stability of frequencies is merely a manifestation of the law of large numbers.

This law plays a very important role in the system of probabilistic concepts (for more details, see Gnedenko, 1969). The law allows us to understand (though in a purely logical aspect) why it is possible to use the theory of probability to solve the problems of the real world. But in no way can it serve as a sufficient reason for justifying broad application of theoretico–probabilistic methods since it is very difficult to give a faultless physical interpretation of the conditions which random values must satisfy in order to obey the law of large numbers (for criticism of the law, see Alimov, 1974).

However, nothing definite can be said about the stability of frequencies in the phenomena of the external world, or about statistical stability in a broader sense. There are many real problems in which statistical stability is precisely the object of research, e.g., in the application of analysis of variance in metrological problems to display the dispersion of results of similar measurements taken by various researchers in various laboratories. It is true, however, that the possibility of applying an analysis of variance stems from certain practically untestable prerequisites.

Probabilistic judgments are built, like any other judgment, by proceeding from certain premises. The grammar of probabilistic statements is, generally speaking, nothing more than the rules of constructing grammatically correct phrases (within a given system of concepts) over initial premises. For example, if we study repeated mass phenomena, we can generate grammatically correct (in our system of concepts) judgments concerning the future. But this extrapolation will be legitimate only if the constancy of frequencies is postulated. Information about the constancy of frequences in the future cannot, generally, be obtained from our past experience, nor can we deduce it from the axiomatics of probability theory. Axiomatics only provides us with a grammar that allows us to state what will happen if we accept certain premises.

Now let us consider a slightly different problem. Assume that we wish to predict the future value of the dependent variable from the observational results, using the equation of the straight line. In this case, we obtain the least-squares estimates of the parameters of the straight line; then we build the limits of confidence as two conjugated hyperbolas and make forecasts for the period we are interested in. But, in doing this, we proceed from the following premises:

(1) Errors of estimating the dependent variable are independent random values sampled from the normally distributed universe with a constant, but unknown, variance and with mathematical expectation equaling zero.

(2) Independent variables are estimated without error.

(3) Both parameters of the regression equation have no time drift.

In this case, the forecast is a proposition correctly constructed over these premises, not all of which are of equal importance; some can be slightly violated. Sometimes we even feel in what way the structure of a phrase must be modified if the premises change. For example, if requirement 2 is not fulfilled, regression analysis is replaced by confluence analysis. The most serious requirement is the third, and it is not quite clear whether it can be included among those pertaining to the concept of "statistical stability." One thing is obvious: either the requirement of "statistical stability" should be regarded very broadly, in which case it cannot be introduced into the language's grammar as a separate category, or it may be regarded in a narrow sense, limited, for example, to premise 1, in which case we shall have to stipulate that probabilistic statements should be based not only upon "statistical stability" in a narrow grammatical sense, but also upon stability in a broad sense, which in various problems is displayed in various ways.

In some applied problems, the requirement of "statistical stability" in its narrow sense is not explicitly formulated at all. As an illustration, let us consider a problem of the science-of-science investigated by my postgraduate student S. A. Zaremba. It deals with studying articles cited according to the years of their publication. It has turned out that at the beginning of the eighteenth century, when science, as an information system, was only in the bud, articles cited were evenly distributed according to the years of their publication. To be more correct, it was a mixed distribution composed of several even distributions given at different sections of the time scale and taken with different weights. In the second half of the eighteenth century, the mixed distribution began to contain an exponential constituent. At first it was situated only at the initial section of the time scale and embraced only a small number of publications. But little by little, as we come closer to our time, the greater is the role of the exponential constituent, though the distribution still remains mixed: its tail part preserves the character of an even distribution. The tail part is at present reduced to several dozens of years, whereas at the beginning of the eighteenth century it went as far back as Aristotle's time. The emergence and evolution of the exponential constituent may be interpreted as the representation of the forefront of development in science, which has no roots in the past (i.e., rapidly attenuates in the reverse time direction). Even distribution may be regarded as a particular (degenerate) case of a truncated exponential distribution, which occurred when publications had a relation to all the past experience.

Everything is thus clear. In this research, the concept of distribution functions was used as a specific language cliché to describe a genuinely complicated phenomenon. And we feel that this phenomenon has been aptly clarified by applying familiar stereotypes of the probabilistic language. Nobody worried about the "statistical stability" here. Distribution functions for adjacent time intervals look alike; those for long intervals look essentially different. This was the object of the research. Here, of course, stability was implicit, allowing one to pass from a single observation of frequencies to the concept of probability.

Constructing concepts of probabilistic language.   There exist a lot of statements about what mathematical statistics is. I find them interesting to collect and have presented my collection (certainly incomplete) in an appendix to my earlier books (Nalimov, 1974b, 1981). In this context it seems pertinent to say that mathematical statistics is a language for constructing statements over values which we like to regard as random.

How was it possible to construct such a language?

Randomness cannot be introduced directly into the system of logical judgments — the latter will immediately prove to be laden with stark contradictions. A system of theoretical constructions had to be formed which would generate concepts enabling the formulation of logically precise descriptions of random phenomena. Among such concepts are general population, sample, probability, distribution function, independent observations, spectral density. These clearly defined ideas and the logical statements built over them are consistent. Randomness has proved to be excluded from the system of logical constructions. It manifests itself only when these constructions are interpreted in the language of experiments, when separate ideas, e.g., mathematical expectation estimated from the sample, are ascribed a fuzzy numerical value, and this fuzzy value is somehow limited by another concept, that of confidence limits. Probability theory, and mathematical statistics in conjunction with it, have reduced the study of randomness to describing random value behavior in probabilistic terms. It has yielded the possibility of describing chance by means of formal logic. The language of such descriptions is weaker than that of causal concepts since it allows us to introduce fuzzy values, at least at the stage of interpretation.

We must be aware of the fact that the concepts of probability theory are certain abstract constructs and not mirror-like reflections of what truly exists in the real world. It is rather a challenge to demonstrate in what way these constructs correspond to what we observe in the real world. There is nothing in the real world to correspond to one of the principal theories of mathematical statistics — that of general population: this concept is a product of profound abstraction. The concept of probability may be shown to correspond to the frequency in the real world if the number of observations is large, though not too large. An overcritical reader will find it hard to understand this. The idea of statistical independence is easily defined in mathematical terms, but it is not so easy to explain to the experimenter how experiments must be carried out in order to get statistically independent results.

On the subject of the difficulty connected with interpreting the term "sample," this is what Tutubalin (1972) says in his brilliant sophism:

> We say that a sample is formed by the results of several independent measurements taken under similar conditions. However, if all experimental conditions are controlled, we shall get one and the same number (there will be no uncertainty), and if not all experimental conditions are controlled, then how do we know that they remain unchanged? (p. 196)

This vagueness of the principal concepts in the sense of their correspondence to reality, of which I could talk much longer, sometimes provocatively gives rise to indignant articles of the kind I have already mentioned (Alimov, 1974). I consider such criticism somewhat illegitimate. One must keep in mind that the language of probabilistic concepts can describe the world only roughly. Let us take the well-known relation

$$\sigma^2\{ \bar{y} \} = \frac{\sigma^2\{y\}}{N}$$

In its practical interpretation this is but an approximate relation, and we never know its degree of approximation. The latter is given, on the one hand, by the fact that real observations can never be absolutely independent; on the other hand, it proceeds from the fact that, N being large, experimental conditions no longer remain constant. I believe an experimenter to be statistically educated if he can use the formula wisely. To be able to use mathematical statistics correctly, one has to interpret the limitations formulated in mathematical language in the language of experiment. But, strictly speaking, nobody knows the rules of interpretation.

The requirements which had to be placed upon the behavior of random values while constructing the principal ideas of probability theory proved fairly rigid. It might seem, perhaps, that the real world is more random than is assumed by the language with which we try to describe randomness. Sometimes these requirements may be weakened. Kolmogorov's frequency interpretation of probability mentioned above (see p. 109) is already a weakened (as compared to von Mises's) idea of statistical stability. Indeed, it is impossible to have an infinite series of tests with experimental conditions held constant. Another weakening of requirements made for the behavior of random values is the introduction of *robust* estimates, i.e., those insensitive to initial premises, instead of Fisher's effective estimates, when measuring distribution parameters from samples. As a matter of fact, a grammar of robust estimates cannot be theoretically constructed, so we have to resort to simulating problems in computers to be able to offer recommendations. However, it is also true that the concept of robust estimates cannot be understood without also understanding the concept of effective estimates. For example, the spectral theory of random processes is built only for stationary processes, whereas all, or almost all, observable processes are non-stationary. If the non-stationary aspect cannot be algorithmically removed, then, as follows from the algorithmic probability theory, the non-stationary aspect itself is random. However, nobody can describe this type of randomness; there are no theories within which it could be described. We deal here with a phenomenon generated by a mechanism more complicated than the algorithms we can construct to describe it. In other words, the algorithm for removing the non-stationary aspect cannot be established more compactly than the random sequence itself. One may, certainly, attempt to describe non-stationary processes in the framework of spectral theory, in the manner of Granger and Hatanaka (1964), but such descriptions will seem clumsy.

Every attempt to weaken the requirements imposed upon the behavior of random values by the grammar of statistics irritates mathematicians dealing with probability theory. Laplace, whose contribution to probability theory is remarkable, remained a convinced determinist. Today, too, mathematicians who deal professionally with probability theory and statistics may still share profound and unyielding formalistic views.

In one of the respectable universities of the USSR, the course of mathematical statistics starts off roughly as follows: ". . . 80% of the applications of statistics are wrong since it is applied where there are no random values." In the book by Tutubalin (1972) cited above, we read:

> It is extremely important to eradicate the delusion, sometimes shared by engineers and naturalists insufficiently trained in probability theory, that any experimental result may be regarded as a random value. (p. 166)

So what are the values considered non-random? Those described by causal relations? This is a fallacy. Tutubalin is quite clear on this point, ascribing to non-random values the results of an experiment for which the requirement of statistical stability is not fulfilled. Non-random is what behaves more randomly than is allowed by the language of traditional probabilistic concepts. Is not this notion of randomness obviously inconsistent with its algorithmic definition?

One cannot say how the requirement of statistical stability should be interpreted in each particular case. To be quite meticulous, one will have to limit the applications of mathematical statistics to such experiments as the tossing of a coin and the applications of probability theory to manipulating balls in urns. Even casting dice is not absolutely random because it is not that easy to make perfect dice.

This is precisely what the *art* of statistical analysis consists of: describing in the language of probabilistic concepts the behavior of the real world which is arranged more randomly than is allowed by the grammar of our language. Such a description is sure to be far from successful in many cases.

When the language of probabilistic concepts proves unfit.   Sometimes statements formulated in the familiar probabilistic language seem clumsy as a result of the fact that the phenomena described actually reflect truly causal relations that are camouflaged. Here is an illustration from Maslov et al. (1963):

> The problem was to give a statistical foundation for measuring dislocations on the ground edge surface of semi-conductor material. The measurements were taken in the following manner: a net was laid over the ground edge, and the number of dots was calculated which

were included in the net cells. The results can be well presented as distribution functions. But the latter proved mixed in this case. Their parametric presentation requires computing high order moments, which is clearly inconvenient since it demands a great number of measurements. Besides, the description in terms of distribution functions proves too cumbersome. This is due to the fact that dots on the surface of the ground edge are situated primarily non-randomly, forming clear-cut figures — stars, spears, or merely clouds of condensation. It has been noted that metallurgical engineers can successfully arrange ground edges according to the quality of the material, immediately connecting the etched figures with the physico-chemical properties of the material. The problem thus turned out to be of an obviously topological character and not of a metrical one: it is not the distance among the dots but their entrance into definite sets forming figures which interests the researcher. When this had become clear, it was suggested that the method of evaluating the material's quality be changed. Laboratory assistants were given albums of real and clearly seen etched figures and they were asked to classify ground edges in accordance with the types represented in the albums. This modified method brought about its own statistical problem: it was necessary to estimate how often laboratory assistants at different times classify the same ground edge as belonging to one and the same type.

This phenomenon reflected a causal (though not too prominent) relation between the etched figure and the material's quality, and it was better not to avoid it by a statistical description.

So who can tell in what cases probabilistic language is to be used and when it is not? No general criterion can be proposed. I think it is applicable when the description obtained with its help satisfies us.

Ontology of Chance

What is the physical nature of chance? It seems impossible to answer this question, at least at present.

I shall remind the reader of the way in which the concept of chance is introduced in mathematical literature. In many books on probability theory (e.g., Gnedenko, 1969), the same phrase is repeated, dating back to Aristotle: an event is called random if, under certain conditions, it may or may not happen. The phrase tells us nothing of the physical sense of the concept. The latter is at times linked with generators of randomness, but any such generator produces, among others, sufficiently well-arranged numerical series.

In his well-known book, Hald (1952) makes an attempt to deduce the notion of randomness from that of stochastic independence:

... A sample of n observations $x_1, x_2, \ldots x_n$, from a population with distribution function $p\{x\}$ is called *a random sample from that population if*

$$p\{x_1, x_2, \ldots, x_n\} = p\{x_1\}p\{x_2\} \ldots p\{x_n\} \qquad (2.1)$$

It follows that $n!$ different possible orders of given sample values are all equally likely when the sampling is random since the value of $p\{x_1, x_2, \ldots, x_n\}$ is independent of permutations of x's when (2.1) is satisfied. A general *definition* of randomness of a sequence of n observations from the same population in terms of the magnitude and order of these observations therefore seems impossible. (p. 338)

Thus, an attempt to define randomness through stochastic independence proves inconsistent with the notion of randomness which follows from the algorithmic theory where randomness is regarded as a maximal disorder.[14]

The notion of random numbers is, actually, an abstraction. In reality we always deal with pseudorandom numbers, and everybody studying the simulation by Monte-Carlo method knows how cautious one should be concerning the randomness of pseudorandom numbers.

Kolmogorov does not explicitly introduce randomness into his axiomatics. His probability theory is constructed in the framework of a general theory of measure with one special assumption: the measure of the whole space must equal unity.

An algorithmic definition of randomness seems to allow a profound comprehension of randomness from a mathematical standpoint, but it hardly elucidates the physical sense of the concept. It is noteworthy that, being interpreted philosophically, the algorithmic approach to randomness is definition by negation: randomness is defined as something which cannot be described in a deterministic way. It is important to comprehend the significance of this statement thoroughly.

If we turn to philosophical literature, we shall again fail to find fruitful considerations of the ontology of chance. In Soviet philosophy, chance is

---

[14] It would be of interest here to pay attention to a paradox of randomness in experimental design problems. It would seem natural to consider the experimental design $\mathbf{X}$ randomly organized if it allows one to obtain stochastically independent estimates of regression coefficients, i.e., such estimates for which cov $\{b_i\, b_j\} = 0$. In this case all the non-diagonal elements of the information matrix $\mathbf{X}^x\mathbf{X}$ should equal zero. But such a design can be built by using, say, a Hadamard matrix. This is a square matrix of the order N consisting of the elements $+1$ and -1 and having the property that $\mathbf{X}^x\mathbf{X} = N\mathbf{I}$. From the definition it follows that all the non-diagonal elements equal zero and, consequently, all the covariances for the regression coefficients estimates also equal zero. If now we try to construct an experimental design of the same dimension randomly placing $+1$ and -1 in the cells of the table, then we, as a rule, obtain designs for which non-diagonal elements will be comparatively small but not equal to zero. It turns out that, at least in some cases, regular modes of construction yield an experimental design generating regression coefficients estimates arranged "more randomly" than designs built randomly. Recently, it became known that to construct a valid sequence of random numbers (satisfying many criteria) one should use not random procedures but some rational ones.

now elevated to the rank of a philosophical category, which, of course, is an important rehabilitation of the concept. But in reality this leads to the following:

> Randomness—a kind of relation determined by external causes secondary for a given phenomenon or a process. Random relations are characterized by unstable and temporal occurrences, relative indifference towards the form of its manifestation, and uncertainty of emergence in space. The category of randomness is correlated to that of necessity. (Yakhot, 1970, p. 33)

I am sure physicists will reject this definition of randomness. Are we to say that the movement of gas particles, experimental errors of measurements, radioactive decay, and probability waves in the microworld are determined by secondary causes? If these secondary causes are removed, experiments will become free from error, radioactive decay will lose its random character, and the concept of probability waves will merely disappear. Mathematicians dealing with probability theory will be even more indignant: they require statistical stability, and the definition says that random relations are characterized by their unstable character.

But let us leave philosophers alone and turn our attention to the literature of popular science. This kind of literature is interesting in that it reflects only what is indubitably acknowledged by the existing paradigm. On the table in front of me there are several books of this kind that have a direct bearing on the matter in question. One of them is the book *This Random, Random, Random World* by Rastrigin (1969). In it we read:

> Indeed, any event has a quite definite cause, that is, is an effect of this cause. Any random event has such a cause, too. (p. 5)
>
> . . . randomness is, first of all, . . . unpredictability resulting from our ignorance, our insufficient knowledge, lack of necessary information. (p. 8)

This sounds like an age-old incantation: "I am not a heretic; I do believe in causality!" However, later on the author has to make concessions and to speak of the uncertainty principle, inexhaustability of the universe and limited human possibilities—in brief, to state the impossibility of getting rid of randomness. So what is randomness, then? Is it only our ignorance and something unique in the microworld?

In the book *Natural Philosophy of Cause and Chance* by Born (1949) we read:

> The notions of cause and chance which I propose to deal with . . . are not specifically physical concepts but have a much wider meaning and application. . . . It would be far beyond my abilities to give an account of all these usages, or to attempt an analysis of the exact significance of the words "cause" and "chance" in each of them. . . .

Indeed, cause expresses the idea of necessity in the relation of events, while chance means just the opposite, complete randomness. Nature, as well as human affairs, seems to be subject to both necessity and accident. Yet even accident is not completely arbitrary, for there are laws of chance, formulated in the mathematical theory of probabili- ty. . . . In fact, if you look through the literature on this problem you will find no satisfactory solution, no general agreement. Only in physics has a systematic attempt been made to use the notions of cause and chance in a way free from contradictions. (p. 1)

Further, Born prefers to speak of the concrete sense the concept of chance has in various physical problems.

We cannot learn what randomness is from the book *Causality and Chance in Modern Physics* by Bohm (1957) either. We read there:

Indeed, the laws of chance are just as necessary as the causal laws themselves.[1] For example, the random character of chance fluctua- tions is, in a wide variety of situations, made inevitable by the ex- tremely complex and manifold character of the external contingen- cies on which the fluctuations depend. . . . Moreover, this random character of the fluctuations is quite often an inherent and indispen- sable part of the normal functioning of many kinds of things, and of their modes of being. (p. 23)

---

[1] Thus, necessity is not to be identified with causality, but is instead a wider category.

Thus, we learn that randomness is inherent to nature and is part of necessity. Probably all this saves us from heresy, making us believe that everything is necessary, but this is hardly essentially elucidating.

In the book by Blokhintsev (1966), His Majesty Chance is introduced without any incantations, and his role in the quantum-mechanical con- ception of the microworld is described. I feel that nothing better can be done.

Any attempts to comprehend the ontology of chance lead to obviously superfluous statements. It seems better to say that randomness is not an ontological cateogry, but an epistemological one. Or that, like necessity, this is one of the two categories generating two languages for describing the world. In both cases, we deal not with concepts emerging as a mirror- like reflection of reality but with abstractions built over the observed ex- ternal world, abstractions generating two different grammars for arrang- ing and comprehending our observational results. Here we cannot but recollect Bohr's principle of complementarity. If this linguistic viewpoint is acknowledged, we immediately succeed in climbing out of the bog of reasonings on chance's ontology and become free from the need to sing incantations. The highly readable collection of papers *Sovremennyi Determinizm* (Svechnikov, 1975) is a fine example of the difficulties one faces in attempting to ascribe an ontological sense to the ideas of causality and chance.

To my mind, the failure of all the attempts to comprehend chance ontologically has a very simple explanation: their aim is to achieve the impossible, i.e., to explain chance in the familiar framework of deterministic ideas.

The idea of the ontological meaning of the concept of chance can be successfully grounded only within the extreme philosophical manifestations of ideas which are customarily referred to as irrationalism. It is noteworthy that the acknowledgment of ontological chance is accompanied by the rejection of ontological causality. I shall illustrate this by a quotation from Sartre's (1965) famous *Nausea*:

> The essential thing is contingency. I mean that, by definition, existence is not necessity. To exist is simply *to be there*; what exists appears, lets itself be *encountered*, but you can never *deduce* it. There are people, I believe, who have understood that. Only they have tried to overcome this contingency by inventing a necessary, causal being. But no necessary being can explain existence: contingency is not an illusion, an appearance which can be dissipated; it is absolute, and consequently perfect gratuitousness. (p. 188)

This elegant statement looks very pertinent in the system of Sartre's existentialism. But here we try to remain within the frame of scientific reasoning.

The description of phenomena in terms of chance makes the world more mysterious than the determinist believes it to be. As a matter of fact, this is a purely psychological effect which disappears during a subsequent logical analysis. Indeed, consistent determinism makes us acknowledge certain initial causes, such as laws of nature which had emerged without cause.[15] Mysteriousness, inherent in determinism, is merely shifted to the remote past. In the system of deterministic ideas, the world, emerging without cause, now proves causally arranged, and probabilistic notions destroy the arrangement and introduce the absence of cause into the description of our everyday experience.

Finally, I would like to show how the phenomena which cannot be described in the framework of causal concepts can be described with the help of the concept of chance. Imagine a physical apparatus registering

---

[15] The statement that initial causes never appeared but had always existed is nothing more than the acknowledgment of difficulties which arise while unwinding the chain of causal links, in a program extrapolating into the past. Actually, it is hard to imagine how something absolutely unchangeable and uncreated which, nevertheless, generates a changing world can exist in time. All this resembles theological structures in which consistent determinism unavoidably leads to the perennial First Cause. But within well-reasoned philosophical–religious systems, e.g., in gnosticism, it was at least stated that the initial cause, God, exists outside time. Moreover, God was sometimes described as "non-existing"; otherwise, one had to look for the cause of his appearance.

the radioactive decay of atoms at some moment of time. What determines the process for a given atom at a given moment of time? Modern physics does not answer this question: one has to acknowledge that the search for the so-called "latent parameters" is clearly a hopeless task (see, for example, Svechnikov, 1975). It may be said that the decay of radioactive atoms obeys statistical regularities. But the sense of this statement lies in the fact that the actually observed frequencies, for some reason or other, behave in such a stable manner that it becomes possible to speak of presenting observational results by distribution functions. The knowledge of their parameters allows us not only to forecast the process of decay in time with great certainty, but also to control many physical experiments. Here we deal with a description which makes it possible to master nature without penetrating into the essence of the phenomena. One can certainly say that statistical regularities are a special case of a broadly understood principle of determinism. Hence, it would seem to follow that, by force of necessity, the given atom does undergo decay at a certain fixed moment of time. However, this statement will hardly differ from the statement that the decay of the atom was caused by the will of the Demiurge, the creator of worlds. In neither case can we support our statements by any substantial data. We cannot acknowledge, even in a purely speculative way, that this decay fixed in time was caused by a trigger, something like an alarm clock randomly set in the infinite past.

Any reasoning of this sort makes things more puzzling rather than clarifying them. Still, this is not to say that we are going to share the viewpoint of agnosticism. We are only made to acknowledge that we have to use a language containing concepts whose physical sense, after serious consideration, proves to be fairly vague. The odd fact is that it is with the help of such concepts that the world is described and mastered. This is an amazing peculiarity of our scientific language. Why should we not discuss these matters directly?

And now a few words about the well-known book *Chance and Necessity* by Monod (1972). In discussing Darwin's evolutionary theory, Monod remarks that the prevailing importance should be attached to the variability at the molecular level rather than to the struggle for existence, an idea that belongs not to Darwin but to Herbert Spencer. However, variability related to the molecular interpretation of Darwinism can be described only in terms of chance: its inexhaustible resources have to be connected, according to Monod, with the ocean of chance. This, in its turn, results in difficulties of comprehension:

> Even today a good many distinguished minds seem unable to accept
> or even to understand that from a source of noise natural selection
> alone and unaided could have drawn all the music of the biosphere.
> (Monod, 1972)

Monod says further that the progress of evolution comes from external conditions that place limitations on chance. But the basic thing is still variability generated by chance. He remarks that not only phylogenesis has to be described in terms of chance, but some local phenomena as well — e.g., the process of formation of specific antibodies for destroying newly emerged antigens. No information is borrowed from antigens when antibodies are synthesized. Everything happens as in playing roulette.

My first reaction to Monod's book was a feeling of sadness. We are so accustomed to perceiving the world in terms of causal concepts that the description in terms of chance seems to lack an explanatory power. However, by and by, a new sensation arose — the impression that biology is now facing a revolution probably even more crucial than that in twentieth century physics. We come to understand that the mystery of life, as well as the mystery of the microworld, can be described only in terms quite new and unfamiliar. The world confronted by modern science proves so complicated that it cannot be described in the familiar system of ideas. To describe this complexity, we had to invent a new language containing concepts with a vague physical sense. I should probably add that the physical meaning of these concepts is unclear because of our desire to comprehend them within the system of old ideas.

Concluding Remarks

Let us try to sum up. Determinism is deeply rooted in the history and pre-history of human thinking. The concept of chance evidently appeared much later, when it was understood that the search for the causal explanation of all phenomena inevitably leads to fantastic conceptions. However, it took a long time to coordinate randomness with a formal, logical way of constructing judgments, and European philosophical thought, both scientific and religious (they were quite in agreement on this question), spent ages to trying to avoid randomness by explaining it simply as insufficient knowledge. Probability theory, having laid serious limitations upon manifestations of randomness, created a language that allowed us to describe the latter within strictly logical structures. This language proved richer than that of rigid determinism and gave the opportunity to describe phenomena in a fuzzy manner without arranging them in a system of rigid causal relations. The position of determinism grew weaker. The most extreme position among probabilistic trends is occupied by the school of subjective probabilities, based upon the neobayesian approach. The prior distribution function can be regarded here as a system of fuzzy (probabilistically weighted) axioms, and the posterior distribution, as a fuzzy judgment. Not only among philosophers, but also among mathematicians who study probability theory, the struggle continues against the acknowledgment of chance: some of the latter are trying to limit probabilistic concepts to an extreme formalism.

Philosophically disposed thinkers like to ask whether there is progress in the history of human thinking. Scientific achievements can hardly be claimed to be the best manifestation of progress. Pragmatically, they certainly have made life easier, but they also have brought mankind face to face with the threat of ecological catastrophe; epistemologically, all scientific results can be interpreted as nothing more than mastering nature, since our knowledge of today, from the point of view of tomorrow, is only paradigmatically fixed ignorance. It is not the change of ideas that matters, but the evolution of human thinking. To this extent, we *have* progressed in our comprehension. One can actually speak only of this progress in thinking. We have to acknowledge that, as science develops, thinking does grow broader. The constricting framework of dulling determinism is collapsing, though from time to time we observe efforts to save it, by implicit or explicit recognition of logical positivism. The acknowledgment of chance is not the only attempt directed at broadening our thinking. Other attempts to manifest the freedom of thinking can be indicated. These include Bohr's principle of complementarity, to which its author endeavored to ascribe a universal character, and attempts to construct a many-valued logic, in particular, the three-valued logic of Reichenbach, designed to formalize physical theories. However, not all of these attempts are welcomed by everyone.

I do not ask the reader to turn to irrationalism. The problem can be solved, at least partially, by weakening formal logic. Without logic we cannot say anything coherent. In my earlier book *In the Labyrinths of Language: A Mathematician's Journey* (Nalimov, 1981), I tried to use the neobayesian approach to explain the irregularity of our verbal behavior, and attentive readers could not but notice that my reasoning was based on common logic. The same rebuke was made by Born to Reichenbach when the latter developed his concept of three-valued logic. Born (1949) wrote:

> Concerning the logical problem itself, I had the impression while reading Reichenbach's book that in explaining three-valued logic he constantly used ordinary logic. This may be avoidable or justifiable. (p. 108).

The same is true of everything written above. It's up to the reader to judge!

Chapter 5

The Distribution Function of Probabilities as a Way to Determine Fuzzy Sets[1]

Sketches for a Metatheory (A Dialogue with Zadeh)

Introduction

The concept of *fuzzy sets* introduced by Zadeh sounded like a challenge to European culture with its dichotomous vision of the World within a strictly discrete system of concepts. In a short period of time, there were many publications devoted to the elaboration of this concept. A bibliography (Gaines and Kohout, 1977) published 12 years after the publication of the initial paper contains 750 titles. In 1978, publication of the journal *Fuzzy Sets and Systems* was started in The Netherlands. The initial paper by Zadeh (1965) was cited 61 times in 1977 according to *Science Citation Index* and 52 times according to *Social Sciences Citation Index*. This is a high level of citation for a paper on applied mathematics. The book by Zadeh (1973b) was published in Russian, and one of his articles was translated for the Soviet serial *Mathematics Today: Mathematics and Cybernetics* (88,930 copies). This seems to testify to the fact that the theory of fuzzy sets has great significance in the Soviet Union. Numerous symposia and conferences on fuzzy sets have been held in many countries.

Being quite aware of the acute formulation of the problem by Zadeh, I sought a precise answer to the question: In what way does the concept of Zadeh logically differ from the probabilistic description of the world which has not yet acquired wide recognition. Zadeh (1978), doing justice to the pioneer papers by Wiener and Shannon, who showed *information* to be internally *statistical*, emphasizes that lately there have emerged

125

problems generally related to the *sense* of information rather than to its transmission as a symbolic system by a channel. He remarks that, in solving these problems, one has to proceed from the *distribution of possibilities* rather than from the probability theory. The distribution of possibilities follows from the conception of fuzzy sets and is, perhaps, the central idea of this theory.

Here I would like to consider critically the legitimacy of comparing Zadeh's theory to the theoretico–probabilistic concepts, especially to those of the subjective probability school.

The Distribution of Possibilities

Zadeh (1978) states that a fuzzy variable **X** is related to the distribution of possibilities Π_x in the same way that a random variable is related to the distribution of probabilities, though in the general case a variable may be related both to the distribution of probabilities and to that of possibilities. According to Zadeh, both of these distributions constitute a weakly linked principle of concordance of *probability/possibility*. This idea is exposed by him in detail as follows (Zadeh, 1978):

> To illustrate the difference between probability and possibility by a simple example, consider the statement "Hans ate X eggs for breakfast," with X taking values in $U = 1, 2, 3, 4, \ldots$ We may associate a possibility distribution with X by interpreting $\pi_x(u)$ as the degree of ease with which Hans can eat u eggs. We may also associate a probability distribution with X by interpreting $P_x(u)$ as the probability of Hans eating u eggs for breakfast. Assuming that we employ some explicit or implicit criterion for assessing the degree of ease with which Hans can eat u eggs for breakfast, the values of $\pi_x(u)$ and $P_x(u)$ might be as shown in Table 1.

> TABLE 1. *The possibility and probability distributions associated with* X

> | u | 1 | 2 | 3 | 4 | 5 | 6 | 7 | 8 |
> |---|---|---|---|---|---|---|---|---|
> | $\pi_x(u)$ | 1 | 1 | 1 | 1 | 0.8 | 0.6 | 0.4 | 0.2 |
> | $P_x(u)$ | 0.1 | 0.8 | 0.1 | 0 | 0 | 0 | 0 | 0 |

We observe that, whereas the possibility that Hans may eat 3 eggs for breakfast is 1, the probability that he may do so might be quite small, e.g., 0.1. Thus, a high degree of possibility does not imply a high degree of probability, nor does a low degree of probability imply a low degree of possibility. However, if an event is impossible, it is bound to be improbable. This heuristic connection between possibilities and probabilities may be stated in the form of what might be called the possibility/probability consistency principle, namely:

If a variable X can take the values $u_1, \ldots, u_n$ with respective possibilities $\Pi = (\pi_1, \ldots, \pi_n)$ and probabilities $P = (p_1, \ldots, p_n)$, then the degree of consistency of the probability distribution P with the possibility distribution Π is expressed by ($+ \underset{=}{\Delta}$ arithmetic sum)

$$\gamma = \pi_1 p_1 + \ldots + \pi_n p_n \qquad (2.13)$$

It should be understood, of course, that the possibility/probability consistency principle is not a precise law or a relationship that is intrinsic in the concepts of possibility and probability. Rather it is an approximate formalization of the heuristic observation that a lessening of the possibility of an event tends to lessen its probability — but not vice-versa. In this sense, the principle is of use in situations in which what is known about a variable X is its possibility — rather than its probability — distribution. In such cases — which occur far more frequently than those in which the reverse is true — the possibility/probability consistency principle provides a basis for the computation of the possibility distribution of the probability distribution of X. Such computations play a particularly important role in decision-making under uncertainty and in the theories of evidence and belief.[1]

---

[1] A. Dempster, Upper and lower probabilities induced by multi-valued mapping. Ann. Math. Statist. 38 (1967) 325–339.

G. Shafer, A Mathematical Theory of Evidence (Princeton University Press, Princeton, NJ, 1976).

E. H. Shortliffe, A model of inexact reasoning in medicine, Math. Biosciences 23 (1975) 351–379.

R. O. Duda, P. F. Hart and N. J. Nilsson, Subjective Bayesian methods for rule-based inference systems, Stanford Research Institute Tech. Note 124, Stanford, CA (1976)

Randomness as a Synonym of Fuzziness

All that was said above on the opposition of the two distributions seems to me a misunderstanding. In a recent paper (Nalimov, 1979), I emphasized the fact that the concept of randomness has an epistemological status rather than an ontological one. At the same time, I stressed that the generally accepted axiomatics (by Kolmogorov) of the probability theory lack the concept of chance. All the manuals contain one and the same phrase, which sounds like an incantation: "A random value is given if its distribution function is given." But the probabilistic distribution function can be considered as a measure of fuzziness of the set on which it is given. However, if such an interpretation is correct, then perhaps randomness is merely a synonym of fuzziness?

Within the European scientific and philosophical tradition randomness was usually opposed to causality; therefore, the former was claimed to be the synonym of ignorance. At present, such a conception of randomness has lost any sense, since knowledge is now often expressed by

Making Ready for Discussion

randomness, i.e., by irreducible fuzziness. For example, in quantum mechanics, Schrödinger's equation

$$i\hbar \, \frac{\partial \psi}{\partial t} \; = \; H\psi$$

looks like a record of a cause–effect relation, though the square of the modulus of the ψ function should be interpreted as a fuzziness of the potential behavior. This interpretation might also include the concept of chance. But in any case, Schrödinger's equation describes our knowledge, not our ignorance. In the physics of elementary particles, the S-matrix (dispersion matrix in the quantum field theory) seems to determine dimension rather than randomness of the microworld.

The problem of a physical interpretation of the concept of probability is rather ticklish. The probability theory, like the rest of mathematics, has developed in the tradition of consistent nominalism, leaving aside the real World. Earlier (Nalimov, 1979), I emphasized that the physical interpretation of probability in no way follows from the axiomatic structure of probability theory. A physical interpretation emerges in the process of solving a concrete problem. We know that statistical studies of mass phenomena involve the frequency interpretations of von Mises and Kolmogorov (without a passage to the limit); in quantum mechanics probability is a measure of the potentiality of an event, and in the Bayesian theory of decision making, probability is a measure of the uncertainty of our assertions. Generalizing, we can say that probability is a measure of the fuzziness of a set of events under study. In such an approach the physical sense of probability will be determined each time by a concrete problem formulation — by the way we structure the field of elementary events.

But if one accepts everything said above, then Zadeh's example describing the behavior of Hans eating eggs for breakfast seems rather odd. This situation contains two fields of elementary events: one is frequential, determined by the number of eggs eaten for breakfast; the other is the field of preferences determining the ease with which Hans eats this or that number of eggs. After normalization (choosing such values of πx that their sum equals unity), the second case will also be the distribution function of probabilities (the rest of the axiomatic requirements are automatically fulfilled). In any case, no criterion can be proposed here which would indicate that after normalizing we deal with two basically different kinds of distribution, requiring a different mathematical approach for their further analysis. The problem of statistical stability will be equally important in both cases. Two different approaches to describing one and the same situation, namely, eating eggs, allow us to construct two distribution functions of probabilities, which will have a different physical interpretation of probability, naturally leading to different statistical decisions.

I find these problems uninteresting because of their artificial nature. In any case, it is clear that the situation of eating eggs does not generate new mathematical (or methodological) problems.

Fuzziness of Language Semantics

Zadeh's idea that the semantic ambiguity of a natural language is an inherent property (Zadeh, 1978) appeals to me. The study of semantic fuzziness is obviously an interesting and very serious task. But again, it is impossible to agree with the assertion by Zadeh (1978) that this problem is possibilistic in its nature rather than probabilistic.

Consider the problem of semantic fuzziness as illustrated by bilingual dictionaries. The fuzziness of *entries* is determined by the set of *explanatory* words. For example, in the two-volume English–Russian dictionary the word "set" is explained by 1,816 words. This semantic fuzziness can be reduced by truncating the periphery of the semantic field: in the small English–Russian dictionary the word "set" is explained by only 96 words. The reduction of the number of explanatory words is made by the compiler of the dictionary on the basis of his personal estimation of the importance of individual semantic fragments of the word's semantic field. A mathematician would ascribe the greatest importance to the meaning of the word "set" which is represented in the combination "a set theory." It is noteworthy that the small English–Russian dictionary lacks the word множество among the explanatory words; instead, we find the words комплект, набор, группа, and круг лиц (the combination "tea set" is translated as чайный сервиз). All these translations hint at the mathematical meaning of the word "set," but to understand this, one must have some mathematical education. By the way, in the large Russian–English dictionary the word множество is translated rather laconically:

> множество // c. great number [greIt . . .]; их было ~ they were
> many; there were lots of them; в ~ in many; in a great number; ~
> хлопот a great deal; или a pack of trouble [. . . trʌbl]

Here the word "set" is not put into correspondence to the Russian word множество. At the same time the German word "Menge" is translated into Russian as множество, and so the Russian term теория множеств proves to be a translation loan word from the German term "Mengenlehre," which for a philologist sounds different than the English "set theory," though the large two-volume English–Russian dictionary among the explanatory words has the word множество, as well as the word combination теория множеств.

All these tiresome and absurd readings, so natural for ordinary languages, are easily explained by a model based on Bayes's theorem that was mentioned in the previous chapter

$$p(\mu|y) \;=\; kp(\mu)p(y|\mu)$$

where $p(\mu)$ is the prior differential distribution function of the meaning of the word μ, which means that each person, on the basis of his personal experience and education, ascribes different weights to different parts of the scale of meanings, and these weights are interpreted as probabilities, since in this case the requirement of normalization is fulfilled. The distribution function $p(y|\mu)$ can be called here a preference function: it is a measure of preference which is given to individual parts of the scale of meanings μ while solving a concrete problem y. In our case, the problem consists of reducing the meaning of the word μ to a small number of explanatory words; k is a normalizing constant, and $p(\mu|y)$ is a posterior distribution of the meaning of the word μ obtained by reducing the prior knowledge along the preference functions that emerge while solving the concrete problem. The compiler of the small English–Russian dictionary received his own personal posterior distribution function $p(\mu|y)$ and then truncated it to conform to the limitations on the number of explanatory words allowed. As a result, the word множество was omitted. If the definition had been written by a compiler with a mathematical background, the word множество would certainly have been included, but the word combination чайный сервиз might have been lost. At least in the case of semantic information, the possibility of reducing the set by truncating it is such an important characteristic of *fuzziness* of this set that it can serve as a *definition* of what a fuzzy set is.

The above example also illustrates the role of normalizing. In real problems, the functions $p(\mu)$ and $p(y|\mu)$ may be such that their maxima will be situated very far from each other on the scale, and if we do not normalize, we can obtain a degenerate function with the abscissas everywhere close to zero. Thus, this case has to be interpreted, as I indicated in the previous chapter, as an emergence of a dark state of consciousness in which a decision cannot be taken. But if we normalize, there will be no interpretational troubles. In particular, this seems to be the case with the probabilistic interpretation of the semantics of jokes. In a joke the preference function $p(y|\mu)$ has its maximum in that part of the scale μ where the values of the function $p(\mu)$ are close to zero, while its small values are placed close to the maximum of $p(\mu)$. Then the function $p(\mu|y)$ proves bimodal: one word acquires two equally important meanings in different, widely separated parts of the semantic scale μ; this looks odd and therefore funny.

In my earlier book (Nalimov, 1974*b*) I showed how the entire variety of our verbal behavior could be explained by using a probabilistic model of language. In the English edition of that book (Nalimov, 1981), as well as in a booklet (Nalimov, 1979), I attempted to make some assertions on the nature of human consciousness based on the same model.

Everything expressed in terms of probabilistic concepts could certainly be described in terms of the distribution of possibilities, and then one could pass from the distribution of possibilities to the distribution of probabilities. But is it necessary to lead the cow onto the roof of the barn in order to feed it with hay? This, of course, is a matter of taste.

Extraparametric (Qualitative) Analysis of the Probabilistic Distribution Function

In mathematical statistics, one can observe a sharp polarization: on the one hand, there exists traditional parametric statistics, and on the other hand, non-parametric statistics. Having once faced the difficulties of giving the distribution function, one feels like dropping iconoclasm and choosing the *middle way* (a term borrowed from ancient Indian philosophy), namely, an extraparametric analysis of the distribution function, somehow resembling the qualitative analysis of differential equations.

The difficulties of constructing the distribution function of probabilities in a Bayesian approach are well known.

1. Even such a seemingly simple operation as normalizing has proved to be rather complicated. If, for example, while estimating the distribution function $p(\mu)$, the expert's square under the curve equals $1 + \alpha$, there are different ways of normalizing: we may be satisfied with dividing by $1 + \alpha$, or our approach may be more sophisticated and, based on some psychological reasons, we may decrease some weight estimates that would result in α equaling zero.

2. It is well known that naive experts tend to construct truncated functions assuming the probabilities of low-probability events to equal zero. This was exactly the case with Zadeh, when he constructed Table 1 for Hans eating eggs. There the distribution function $P_x(u)$ is truncated after $u = 3$. In such a truncated version of the distribution function, the application of the Bayesian approach becomes absurd, since it is evident that in extraordinary situations the preference function will reach its maximum under such values of u, where $P_x(u) = 0$, and this preference will be destroyed by the prior limitation. Determining $P_x(u)$ the way it is done in Table 1, Zadeh assumes that Hans, even being extremely hungry, will never eat more than three eggs, though from the numerical values of $\pi_x(u)$ it follows that to eat four eggs for Hans is no more difficult than to eat one egg. Therefore, the possibility of eating more than three eggs cannot be excluded; this event should be ascribed a low probability but never a zero probability. Such illogicality seems to have some psychological basis: an expert usually tends to give biased estimates.

3. Especially great and sometimes insurmountable difficulties often arise during the construction of a space of elementary events and the determination of its metrics. [Recall that *metrics* is a way of determining the distance between two elements (points) or determining the measure of an angle in a geometric system.] Actually, it is not quite clear how one can construct a space of elementary events for language semantics. Generally speaking, semantics does not yield to metricization. All of the publications in which Bayesian statistics is used are confined to problems with well-metricized variables, e.g., those of quality control, etc. This accounts for meager application of the Bayesian approach to psychology and linguistics.

The difficulties enumerated in point 3 can be overcome only if we radically change our attitude toward Bayesian analysis. We shall have to reject the parametric analysis of the probabilistic distribution function, limiting ourselves to *qualitative* considerations. The curve of the distribution function proves mathematically *inexpressible*, but *presentable*.

Here we must involuntarily take the position of nominalism, which is, by the way, only natural since a model is but a metaphor which behaves similarly to the phenomenon described but not exactly the same. (In English the metaphor is well described by the word combination "as if but not.") In the case of language semantics, the train of thought may be as follows: we represent human verbal behavior as if the mind contains a semantic metric on which a probabilistic distribution function is given. I do not know how this metric is given and do not choose to speculate on this subject. I consider it more reasonable to do otherwise: to consider, hypothetically, various forms (shapes) of the distribution curves fit to describe the situation under study, and to obtain propositions explaining the situation without resorting to any numerical values. The Bayesian model thus turns into a logical formula used to infer an assertion from two probabilistic, fuzzy premises. This may be considered as a probabilistic analogue of the syllogism of bimodal logic. Our reasoning always goes along the following route; describing a concrete situation, we consider hypothetically possible forms of the function $p(\mu)$ and $p(y|\mu)$ and then, using Bayes's theorem, we obtain the function $p(\mu|y)$, which helps to explain the phenomenon observed. The fact that we can consider a lot of phenomena within such a system of assertions creates the impression of generality. This also opens up the possibility of constructing hypothetical assertions of a higher hierarchical level. We have passed from the analysis of human verbal behavior to discussing the problem of human consciousness (Nalimov, 1979).

An example of Bayesian logic was given above, when I spoke of the semantics of the word "set" in the dictionary. Broad applications of this logic can be found in my earlier books (Nalimov, 1974*b*, 1979, 1981). In an article written with Meyen, an attempt was made to give a probabilistically weighted, fuzzy description of the World (Meyen and Nalimov, 1978).

I would like to emphasize the fact that mathematical models usually have one of the two possible statuses: logical or numerical. Probabilistic–statistical models usually have a numerical status: their verification (or, according to Popper, falsification) is realized by numerical comparison with empirical data. In contrast, models such as those on the qualitative theory of differential equations are only rarely elaborated to an extent that allows their numerical comparison with the experiment. In this case, we deal with a purely logical construction: the dynamic properties of the system under study are obtained in a purely mathematical way from the hypothetical system of differential equations. The legitimacy of the initial premises is determined by the meaningfulness of their consequences. In our case, something similar happens: leaving the tradition of probabilistic and statistical simulation, we reject all the numerical comparisons and suggest that the legitimacy of our results be evaluated by their speculative non-triviality.

Now let us return to Zadeh. In constructing the possibilistic distribution function, we unavoidably have to cope with the difficulties of point 3 since the problem proves to be *metric* and in no way topological, and Zadeh solves it until it is numerically completed. How this helps to solve the difficulties of metricizing semantics remains obscure. In the papers where Zadeh introduces the notion of a linguistic variable, the significance of this difficulty does not become apparent because he considers only extremely simple situations which could not interest a linguist.

Here we have to deal with another difficulty: experts' estimates are always unstable. And while this instability is natural for the probabilistic–statistical approach, Zadeh should have eliminated it in his constructions. However, it is practically impossible to do so. Followers of Zadeh thus face the necessity of something like the analysis of variance and of evaluating the divergence in the estimates of one expert at different moments of time and under different conditions, then the divergence between different experts, etc. This would unavoidably bring forth the necessity of selecting criteria, and the problem would necessarily become statistical.

At the initial stage of my linguo–semantic studies, I expended great efforts trying to obtain the prior distribution function $p(\mu)$ for a list of words in real psychological experiments. Eventually, I was forced to conclude that these attempts were futile: a person does not wish to, or perhaps cannot, reveal in routine experiments the deep processes of his mind which occur during the comprehension of verbal texts. These processes are too intimate. It was necessary to apply autogenous training prior to the experiments by means of group meditation over the meaning of certain words. Only then did we manage to obtain the fuzzy semantic fields underlying words, the discrete symbols of our language. The subjects allowed access to their deep, intimate processes after their minds had been liberated from the logical structures determined by our culture. We experimented among different groups of people in different places: in the Armenian mountains and in the center of Moscow. The results were always similar. (Different results were obtained only in experiments with patients in a hospital for mental diseases.) Experimental results took the form both of verbal texts and of painting, when our subjects were professional painters. In both cases, semantic fields proved to have been generated by the images having an archetypic nature. [Preliminary results of these experiments have been published (Nalimov, Kuznetsov, and Drogalina, 1978).] These images, when properly interpreted, helped to reveal the semantics of the words under study as a special state of consciousness which people enter when they try to comprehend the meaning of a verbal text. From all said above, despite its brevity, it follows that our probabilistic model of language only hints at what happens in the mind during the interpretation of the meaning of verbal texts. We could hardly expect anything more. Though the model cannot adequately reflect the actual phenomena, it proves sufficient to outline a new direction of research, and for this reason I am satisfied with the quantitative analysis of the model.

However, confining ourselves to the qualitative analysis, we naturally proceed from the assumption that semantic metrics is roughly the same for all sane people of the given cultural stratum. Psychic diseases destroy such structures: it is known that schizophrenics do not understand jokes (Nalimov, 1974b, 1981). It is likely that in this case the semantic scale μ narrows down so that no bimodal posterior distribution function of the word meaning can be obtained.

Concluding Remarks

I am quite aware of the possibility of using a variety of languages to describe a single phenomenon. If we choose one language, e.g, that of *fuzzy sets*, then many dialects of one language are allowed to exist. In order to compare languages (or dialects of one language), it is necessary to construct a metalanguage. This task is also hopeless (since it is possible to construct a lot of metalanguages as well), though I have made an attempt herein to outline a metatheory for the probabilistic approach to the theory of fuzzy sets. It is better to evaluate the advantages of different languages or dialects on the basis of the direct (unformalized) comparison of texts written in these languages. Our works on language (Nalimov, 1979), thinking (Nalimov, 1974b, 1981), and the fuzzy ontology of the world (Meyen and Nalimov, 1978) written in the probabilistic language can be contrasted to the works of Zadeh (1972, 1973a, b) in which he introduces the notion of a "linguistic variable."

It is up to the reader to choose between these two dialects of fuzzy sets. It is noteworthy that Zadeh's approach has an important advantage: he suggests the construction of a completely new language, and thus provides jobs for mathematicians.

In concluding this chapter, I ask: Is not the probabilistically weighted vision of the World a realization of the dream of Pythagoras and Plotinus of describing the World in its integrity and fuzziness through numbers? Note what Plotinus wrote on this point (MacKenna, 1957):

> 1.   A thing, in fact becomes a manifold when, unable to remain self-centered, it flows outward and by that dissipation takes extension: utterly losing unity it becomes a manifold, since there is nothing to bind part to part; when, with all this overflowing, it becomes something definite, there is a magnitude.
>
> 2.   . . . we make a man a multiple by counting up his various characteristics, his beauty and the rest . . .
>
> 3.   Whatever is an actual existence is by that very fact determined numerically . . . approach the thing as a unit and you find it manifold; call it a manifold, and again you falsify, for when the single thing is not a unity neither is the total a manifold . . . Thus it is not true to speak of it [matter, the unlimited] as being solely in flux.
>
> 7.   It is inevitably necessary to think of all as contained within one nature; one nature must hold and encompass all; . . . But within the unity There, the several entities have each its own distinct existence.
>
> 10.   When it takes lot with multiplicity, Being becomes Number by the fact of awakening to manifoldness;
>
> 13.   If, then, unity is more pronounced in the continuous, and more again where there is no separation by part, this is clearly because there exists, in real existence, something which is a Nature or Principle of Unity.
>
> 14.   . . . while continuous quantity exists, discrete quantity does not — and this though continuous quantity is measured by the discrete.
>
> 15.   . . . the Intellectual-Principle, its moral wisdom, its virtues, its knowledge, all whose possession makes that Principle what it is.
>
> 16.   . . . The number belonging to body is an essence of the order of body; the number belonging to Soul constitutes the essence of souls.
>
> 18.   It appears then that Number in that realm is definite . . . There every being is measure; and therefore it is that all is beautiful.

Throughout the entire history of philosophy, the tractatus *On Numbers* by Plotinus was considered a most mysterious and ambiguous work. But now its meaning is suddenly revealed: this is a vision of the World in its fuzziness determined by the distribution of probabilities, by the measure which we still cannot express numerically since we have not mastered the metrics of the World. But has the World any metrics at all? It seems more reasonable to speak of the metrics of spaces of human semantics. The alteration of this metric may occur as a result of both personal (individual) peculiarities and cultural evolution; this is the change in the vision of the World. The creation of a universal language is a dream which always escapes us.

So perhaps the fuzzy, probabilistically weighted vision of the World is the restoration of Ancient Greek numerical dialectics.

Chapter 6

On Some Parallels Between the Bohr Complementarity Principle and the Metaphoric Structure of Ordinary Language[1]

A psychiatrist asks a patient to explain the meaning of the expression "hands of gold." The patient answers: "Well, after amputation one has hands made of gold."

In our ordinary language, there are two logical sequences of meaning that converge into a metaphor. In the above example, one sequence deals with the notion of gold as precious metal and the other, with a notion of a skilled craftsman capable of making wonderful things. The person whose psyche is injured perceives only the first sequence. The inability to recognize metaphors is one of the diagnostic signs of psychic disease (Kasanin, 1944).

Scientific language also has elements of metaphoric character. This has been discussed in the literature more than once [e.g., by McCormac (1971) and also in the chapter "The Language of Science" in my earlier book (Nalimov, 1981)]. Let us discuss here only one example dealing with the term "metascience" and its derivatives. The *Philosophical Dictionary* (Rosental, 1972) says:

> *Meta–Galaxy* (literally: "something beyond the Galaxy") — a cosmic system including milliards of Galaxies . . .
>
> *Metalogic* — a theory studying systems and concepts (metatheory) of the modern formal logic.

[1] This chapter was published in the collection of papers *Complementarity Principle and Materialistic Dialectics* (Nauka, Moscow, 1976). It was translated by L. R. Moshinskaya.

Metamathematics (a proof theory)—a theory which deals with the study of various properties of formal systems and calculus (consistency, completeness, etc.) . . .

Metatheory—a theory which has some other theories as a subject. Metatheory studies a system of propositions and concepts of the given theory; states its boundaries, means of the introduction of new concepts and proof, etc., allowing the possibility for more rational means of constructing a theory . . .

Metaphysics—I. The term "metaphysics" emerged in the first century B.C. as a name for a certain part of Aristotle's philosophical heritage and its literal meaning is "that which is behind physics." Aristotle himself called this most important, in his opinion, field of his philosophical teaching "the prime philosophy" investigating some superior, unavailable to the perceptive organs, speculative truths and unchangeable beginnings of the whole of existence, which are necessary for all sciences. In this meaning the term "metaphysics" was used in subsequent philosophy. In medieval philosophy, metaphysics served theology as its philosophical foundation. . . . II. In modern times, there emerged a depiction of metaphysics as an antidialectical means of thinking, as a result of onesidedness in cognition, when things and phenomena are considered unchanging and independent from one another, and wherein the inner contradictions are rejected as a source for development in nature and society . . .

But the meaning of the term metaphysics is not restricted to the above-given definitions. The XVth International Philosophical Congress in Varna in 1973, its activity strangely influenced by Marxist philosophy, included a session entitled "Modern Metaphysics." The following are examples of the reports presented to this session: "Metaphysik als Nihilismus" by A. G. Bucher (German Democratic Republic); "Metaphysics and Science" by G. Leclerc (United States); "Once Again: Can Metaphysics Be a Science?" by E. Panova (Bulgaria); "Cosmos" by M. Nicolodi (Italy); and "Ontological Content and Epistemological Functions of Philosophical Categories" by V. V. Il'in (USSR).

So we see that the term metaphysics proves to be deeply metaphoric: it ties up more than two lines of logical thought.

Now let us dwell on the metaphoric meaning of the term metamathematics. Metamathematics includes Gödel's well-known theorem on undecidability. The structure of this theorem is the same as that of an ordinary mathematical problem; actually, it is in this way that mathematicians regard it. But at the same time, this work has a deep philosophical meaning; perhaps it is the strongest of all results that have been obtained in epistemology so far. It follows from this theorem that human thinking is far richer than its deductive form and that we cannot build an artificial intelligence on the basis of formal logic. All this is carefully considered in a small book by Nagel and Newman (1960). Gödel's theorem allows us to understand the nature of the difficulties faced by physics in its attempts to create too strict a formalization of its theories. Here the difficulties seem to lie in the nature of our thought rather than in the actual nature of the external world. This question still remains insufficiently studied, though the work *Gödel and Physical Theory* by Yourgrau (1969) is exceedingly interesting to read.

Thus, we see that the term "metamathematics" unites at least two logical lines of thought, and if one perceives only one of them, it causes at least surprise. If, say, a student of mathematics says on an examination that Gödel's theorem lies outside science since it belongs to metamathematics, he will, most probably, get a low mark. But in a paper by Livshitz (1973), with the pretentious title "The Solution Which Does Not Correspond to a Serious Problem," furious attacks are made against the attempt at interpreting mathematical science as metascience. The following words are used against the concept of "metascience": "According to the strict meaning of the term (and also analogous to the Aristotelian "metaphysics") a metascience must mean something found behind, existing *outside, out of the boundaries of ordinary science.*" Does all this not recall the situation described at the beginning of this chapter?

If human thought is really richer than its deductive form, then human language must have the means to express this variety. I have already stated more than once the idea that ambiguity of language, its polymorphism, is a means for overcoming Gödel's difficulty in the logical structure of our speech behavior (Nalimov and Mul'chenko, 1972). Metaphor is just one of the means which facilitates the manifestation of this polymorphism (Nalimov, 1974b, 1981). From the point of view of formal logic, this is the rejection of one of the main logical laws: the law of the excluded middle. Any statement is either truthful or untruthful; the middle is excluded. If we consider a statement about the identity of the objects A and B, then the law of the excluded middle reads as follows: A is either B, or non-B.

Let us turn at last to Bohr's complementarity principle (see, for example, Bazhenov, 1976). According to this principle, in the process of world description, to reproduce the integrity of an object, it is necessary to apply mutually exclusive "additional" classes of ideas, each generating its own logically consistent line of judgments but still proving logically incompatible with the others. Is this merely the transfer of the metaphorical principle of judgment construction in the ordinary language into the language of physical theories? The complementarity principle, as I see it,

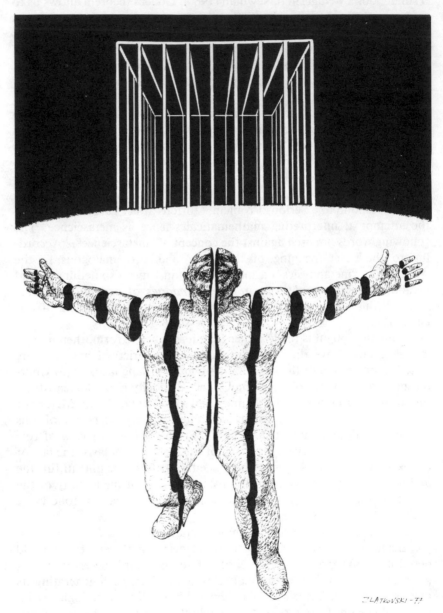

The Complementarity Principle

is just a further widening or, perhaps, strengthening of the method which is daily used in our speech behavior, though on a narrower scale. We are accustomed to metaphors; that is why we seemed to be prepared psychologically to understand Bohr's complementarity principle more readily than, say, Reichenbach's two-digit–three-digit logic.

Bohr (1955) himself gave general philosophical importance to his complementarity principle:

> . . . the integrity of living organisms and characteristics of Homo sapiens, and also of human cultures represent features of the integrity, for which reflexion demands typically complementary means of description.

In my opinion, this is a broadening of the possibilities of formal logic which were discussed in Chapter 4. Let me stress again that there is no passing to irrationalism: formal logic is not thrown away; it is merely used in a restricted sense. In describing the totality of a system, we are allowed to violate one of the principal laws of logic.

Chapter 7

Science and the Biosphere

An Attempt at a Comparative Study of the Two Systems[1]

Introduction

I shall now attempt to consider science as a self-organized system or, still better, as a macroorganism, developing according to a certain set of rules. These rules themselves are developing and changing as the macroorganism develops. In such a study it seems natural to compare science with another well-known macroorganism, that is, with the Earth's biosphere. We need such a comparative analysis only to understand better the unique qualities of the organization of science. It is possible that this comparative analysis will enable us to perceive certain mechanisms of biological evolution. Surely the development of science, its history and pre-history, is easier to understand than the similar processes in biological evolution. In any case, if we succeed in comprehending various aspects of the functioning of one of the systems, it will make us consider the other one from a new standpoint, and herein lies the heuristic sense of this approach.

My attention was drawn to such a comparison by a paper by Shreider and Osipova (1969). While considering the question of the growth of science, they made a number of comparisons with biological evolution, referring to a very unusual description of the evolutionary process in biology given by Teilhard de Chardin (1965).

---

[1] This chapter was written with Z. M. Mul'chenko. The major part of it was published in Russian in *Priroda* (no. 11, 1970).

145

Science and the Biosphere

There is a remarkable characteristic that is typical both of science and of the biosphere. Both these macroorganisms represent systems which are developing in time, so that the information which they contain gets renewed and at the same time becomes more complicated. As the process of evolution continues in the biosphere, new species are coming into being. In science, new concepts and new fields of knowledge arise. Such systems may be called informationally developing. Other systems are also capable of developing, e.g., a growing crystal or plasma, but they are not informationally developing. In the course of their development, essentially new, unpredictable phenomena do not arise which could influence the possible development of the system. To put it briefly, on a large time scale such systems do not evolve.

The Emergence of Systems

The development of the biosphere as a macroorganism began with the emergence of the cell (this idea is expressed by Teilhard de Chardin). The life of the cell and its further development goes on according to a certain set of rules; these can be easily systematized though this work has not yet been completed by biologists (Schmalgauzen, 1968). The biosphere as a system, or a macroorganism, came into existence only after the emergence of the cell. Actually, biologists know next to nothing about the pre-cell era in the biosphere.

Science appeared in a similar way. Science as a self-organizing system emerged with the appearance of the modern system of scientific communication (with such elements as journals and publications).[2] The latter are elementary carriers of information, and they play the role played by a cell in the biological system. There are rules which regulate the emergence of publications and their interrelation. I shall discuss these rules later, and for the present only remark that in science, unlike in biology, we do know the pre-systematic state. We know that in different historic periods there appeared separate flashes of science, but they did not create a self-organizing information system and soon faded. The question is often posed: What actually happened to science in the seventeenth century? Why was it then that the exponential explosion began, which is still going on today? The formal answer to the question is easy: It was then that the information system which turned science into a self-organizing system was being created. But this answer is almost pointless. As a matter of fact, we know next to nothing about the reasons and the way this information system came into existence, nor can we explain why it had not emerged before. It would be very interesting and important to make a profound study of the history and pre-history of the emergence of science as a self-organizing system with its specific network of internal communications.

The process of the emergence of the modern communication system in science can be cursorily traced back, for example, in the library of an ancient university. One such Soviet university is the University of Vilnius. Its library contains scientific literature of the eighteenth and even of the seventeenth century. Looking through this literature in a successive order, we can follow the way this system was coming into existence. But to see means nothing; it is necessary to study the phenomenon.[3]

Would such an investigation be of any help to biologists–evolutionists? Would they be able to see something which is not accessible to their eyes: namely, the process of development of a self-organizing system?

Here again I would like to make a comparison with such systems as plasma or a crystal. External factors create the unique variety of crystals formed by the defects of growth, i.e., dislocations. However, this variety of forms is not retained in a crystal's memory. In this case there is no information system to permit the storage of some features and then the strengthening or weakening of them in interaction with the external world.

---

[2] By "publication" we mean a scientific subject, prepared as a paper, preprint, report, etc. These were often published by new scientific societies such as the Royal Society of London and similar societies in Paris and Berlin.

[3] First steps in this direction have been taken by S. A. Zaremba, a postgraduate student of mine, who has undertaken a study of the evolution of the scientific citation system beginning with the eighteenth century. Her data have not been published as yet.

Informationally Developing Systems Should Have Their Own Mechanism to Overcome Gödel's Difficulties

The development of science is predetermined by discoveries. Discoveries, by definition, are not predictable; they cannot be derived by means of deductive logic from a certain finite set of initial premises. This question has already been discussed elsewhere (Nalimov and Mul'chenko, 1969). The development of the biosphere to its present stage cannot be thought of as a process going on according to some algorithm constructed on principles of deductive logic. To clear up this idea, I shall fall back on Gödel's undecidability theorem, which demonstrates the limitation of logical systems. I use the term "a logical system" to denote a certain set of axioms and the derivation rules. The axioms form a string of symbols, and by repeated application of the derivation rules to this string, new strings can be obtained; that is, theorems can be proved. It follows from Gödel's theorem that, if the derivation rules are finite and deterministic, then any sufficiently rich logical system is incomplete. There exist true statements which can be expressed in the language of this system, but which are impossible to prove within such a system. It is impossible to prove the consistency of such a system by means expressible in the same system. No indefinitely large but strictly fixed increase of axioms can make the system complete (Arbib, 1964; Nagel and Newman, 1960).

Consequently, an informationally developing system cannot be strong enough if it is built according to an algorithm based on deductive logic. We cannot think of biological evolution as a particular system consisting of some premises and finite and strictly deterministic derivation rules which could be interpreted as absolute laws of nature. Similarly, science cannot be regarded as a system which is developing in a strictly deterministic way proceeding from a certain system of postulates.

A Model with a Chance Generator

In what way, then, is Gödel's difficulty overcome in the development both of science and of the biological system? There are two possible explanations. The first is to assume the existence of some mechanism incomprehensible for us in principle; the second explanation is to construct a probabilistic model with a chance generator. In the case of biological evolution we may speak of some incomprehensible vitalistic power, or free will. The hypothesis about the random character of mutations will be an alternative to this statement.[4] Physically, the chance generator is

---

[4] The French biologist Monod (1972) writes: "Pure chance, absolutely free, but blind, lies at the very root of the stupendous edifice of evolution. This central concept of modern biology is no longer one

realized in the form of hard radiation, acting on the genes. Further, we shall have to assume the existence of some system of the selection rules which allow survival only of the individuals with useful (to a certain extent) combinations of features. The chance generator, together with the selection rules, creates an adaptation system. Evidently, adaptability is a peculiar feature of all the informationally developing systems, including biological ones. The difference between animate and inanimate nature lies, first of all, in the adaptability of the former. Inanimate systems—a growing crystal or plasma—are practically unable to adapt. Another characteristic of an informationally developing system is its memory: the system stores novel and favorable information. An interesting idea was suggested by Quastler (1964): an arrival from another planet would not be able to distinguish the system constructed in this manner from a system produced by "free will."

The state of events in the development of science is approximately the same. Here we can also try to build a probabilistic model based on the assumption that the thinking of creative scientists is directed by the chance generator in combination with a very complicated system of the selection rules. Perhaps the intellectual behavior of men of genius is determined by a system of selection rules so perfectly arranged that it allows them to reject unfavorable ideas. The hypothesis of a vague mechanism of metalogical thinking is an alternative to this probabilistic model. We can assume the existence of a hierarchy of the forms of thinking: pre-logical, i.e., imaginative, thinking; logical thinking, i.e., deductive logic; and metalogical thinking. But this hypothesis does not eliminate Gödel's difficulty. It is most likely that scientists at different levels of activity are in different stages of the hierarchy of thinking, but they communicate with one another always at the same level—the logical level. Here we encounter Gödel's difficulty, but it may be overcome by the polymorphism of ordinary language. Different meanings are ascribed to words, and this allows scientists to be not entirely logical. This lack of logic appears to be softened by the polymorphism of the language; it does not jar the interlocutor's ear. It is noteworthy that polymorphism of language not only widens the axiomatic system but also introduces into it some internal contradictions and that the system, strictly speaking, is no longer "Gödelian."

Sometimes we have to pay dearly for the polymorphism of language. Long scientific arguments often arise simply as a result of the fact that the same words are given different meanings, though perhaps it is this which forms an inevitable component of the creative process. Here again

---

among other possible or even conceivable hypotheses, the only one that squares with observed and tested facts. And nothing warrants the supposition—or the hope—that on this score our position is likely ever to be revised."

we can draw a parallel between science and biology, where an error in code reading may cause mutations.

Incidentally, the polymorphic nature of language is one of the difficulties encountered in constructing artifical intelligence such as a computer participating in a dialogue; too strict limitations have to be imposed on the rules which the human being must follow in his dialogue with the computer.

In principle, the model of creative thinking with the chance generator and the selection rules can be verified. Such a verification will undoubtedly be carried out in the process of creating artificial intelligence. At present, it appears to be difficult to construct a model of artificial intelligence which would not include, to some extent, the chance generator. Much effort is being spent on solving the problem of artificial intelligence, especially in the United States.

Here it is noteworthy that, in an attempt to construct a model of language that reflects both its hard (logical) structure and its soft component (the general irregularity connected with the polymorphism of words), we again face the probabilistic model of language discussed in my earlier books (Nalimov, 1974b, 1981).

The Language of Science and the Biosphere, Two Self-Organizing Systems

Earlier, I stated that both science and the biosphere are self-organizing information systems,[5] controlled by their own information flows. The external conditions are the medium in which these systems are developing. This environment may be either favorable or unfavorable for the development of the system, but the environment cannot make the system develop along an alien path. The emergence of new ideas or new forms in the system is given by its information flows. The state of science in Nazi Germany serves as an excellent example. The external environment tried to force science to develop along an alien path, and certain investigations that fit the Nazi spirit were supported by enormous sums of money. However, nothing came of it—a new Nazi science was not created.

An information system should have its own specific language. In biology there is the recently deciphered language of the genetic code, or perhaps in more general terms the language of intermolecular interactions.[6]

---

[5] Here we do not consider the whole hierarchy of systems. Science is a subsystem of the sociocultural system, etc.

[6] When speaking of a language, we should foresee the possibility of building hierarchical systems within language. One of the peculiarities of language is that it can be represented in different sign systems, which form a hierarchical system of different levels. For example, Russian has a system of levels which consists of letters, morphemes, phrases, etc. The same hierarchical structure is found in the language of a biological system at all its levels. But this is a subject for special discussion.

We can say that before our eyes the lines of biological grammar are coming into being (Chargaff, 1971). Science has its specific, though inadequately studied, language of scientific communications, which differs essentially from our ordinary colloquial language. In the process of science's development, its language becomes more and more similar to a code system in which the content of every coding sign is constantly on the increase. As a result of this evolution of language, scientific articles become more compact, though this natural evolutionary process still does not advance quickly enough. Its failure to keep up with the exponential growth of publications has resulted in the information crisis in science. The tendency for an increase of codification capacity moves the language of science farther and farther away from the ordinary language of most people and brings it nearer to the language of the genetic code.

Several examples illustrate the code-like character of the language of science. First of all is the associative language of bibliographic citations. One citation may code a complicated concept stated in a previous publication. Another example is what we may call specific languages of science. With the growth and differentiation of science, closed fields of knowledge with their specific languages are created. The creation of such specific languages makes the exchange of information easier within the limits of separate narrow disciplines but more difficult at the level of the whole of scientific research. Another comparison with biology seems appropriate here: information structures of different species arc incompatible. The two-dimensional language of chemical formulas is an example of a very capacious code language in science. A final example is the mathematization of knowledge, the creation of mathematical metalanguages of science. All this means an ever-growing tendency to codification and to an increase in the capacity of the coding signs.

The differentiation of science is supported by the creation of specific languages, and thus in science there appears "the Babelian problem" (for greater detail, see Nalimov, 1981). In the biosphere, "interspecies difference" within the information systems excludes "interspecies hybridization." Each evolutionary branch stews in its own juice, without information interaction with neighboring branches.

It is noteworthy that even A. Schleicher (1821–1868), whose views were called naturalistic or, in modern terms, precybernetic, regarded language as a natural organism. His own words run as follows (Schleicher, 1888): "The life of language does not essentially differ from the life of any other organism—plants or animals." More recently, Shaumyan (1977) wrote along the same lines:

> Objective grammar [logical mechanism encoded in the speaker's mind] has a peculiar ontologic status: on the one hand, it exists only in human consciousness, and on the other hand, a human being is compelled to treat it as an independent external object. Objective

grammars belong to a specific world which may be called a world of semiotic systems, or a semiotic world. The specificity of the semiotic world lies in the fact that originally it is the product of human consciousness. If by "the materiality of the world" we mean its belonging to objective reality, then the semiotic world as well as the physical one must be also considered materialistic.

For a long time, the human languages and the languages of the biological code developed independently, or almost independently, of each other. One of them dealt with the biological sphere; the other, with the social one. At present, we can speak about a strong interaction between these originally heterogeneous languages, or, in other words, about the interaction between the two worlds—biological and semiotic. The biological language has proved incapable of struggling for existence in these new conditions. Hence, the ecological crisis and the death of many species of animals and plants—the bearers of separate texts written in the language of biological code—which results in the possible death of the language itself. The struggle for existence of languages is a struggle for survival of the texts written in those languages.

Interaction with External Information

Both in the biosphere and in science there is a continuous interaction between newly generated ideas and information received from the external world. In the biosphere, new external information results from fluctuations of factors which determine the external living conditions. In science, new external information arises as a result of new experiments. In both cases, we may speak of the pressure exerted upon the internal information flows of the system by the external information. Using the approach of M. Mednikov, we shall consider the following variants.

1. The external information grows rapidly and, at the same time, the possibility of generating new internal information remains at a high level or, as biologists would say, a high capability, for the inherited variability is retained. This is the most favorable case both in the biosphere and in science. In the biosphere, a new species will appear which will be able to survive in the changed conditions, and in science, a new theory will arise to explain new, experimentally observed facts.

2. The external information remains unchanged; the internal information tends to grow. In the biosphere, we find original, sometimes grotesque forms. Likewise, science degenerates into a system of original, but purely artificial, constructions, as, for example, sometimes happens in philosophy.

3. The external information grows rapidly; the internal, slowly. In the

biosphere this causes the dying out of a species, and in science it leads to difficulties like those we see now in the physics of elementary particles.

4. There is a lack of some factor in the environment, e.g., the light for the subterranean animals. It immediately causes the regression of the species. If an experiment is lost to science, it likewise causes regressions. [Mathematics does not fit this system. It occupies a peculiar position in the scale of sciences. It is not without reason that R. Feynman states that mathematics cannot be considered a science because, unlike other fields of science, it is not based on experimentation (see Schmalgauzen, 1968).]

Exponential and Logistic Growth: Creation of Favorable Ecological Situations

In biology some species, at least at the initial stage of their development, increase in number almost exponentially, and later, as external sources are exhausted, the exponential curve is transformed into a logistic or some other S-shaped curve with saturation. Observations show that a biological system is constructed in such a way that it transforms the Earth favorably for itself, improving the ecological situation. This process can be observed directly during a trip in the mountains, where we can see the plants slowly destroying the rocks.

Something similar takes place in science. Publications, journals, and the number of scientists all, roughly speaking, grow exponentially (Price, 1963) provided there are sufficient resources. At the same time, science as a system creates the ecological situation favorable for itself. Under the influence of science, engineering develops as well, and this creates the means necessary for the development of science by releasing manpower for work in science and creating the industry to manufacture scientific apparatus. True, there is a threat of crisis in this field: the expenditure on science may grow more rapidly than the national product and the state budget.

Naturally, science could have begun to create a favorable ecological situation only under certain economic, social, and political structures of society. It is here that the historical theories about the appearance of science as a system must manifest themselves.

Fecundation as a Process of Information Interaction

As a rule, any biological system is bisexual. The merging of two cells can be considered as a special form of information interaction of the elementary carriers of biological information. First of all, this is necessary for the correction of possible genetic defects of one of the cells. Further, it is also necessary for enriching the daughter species with new characteristics. This enrichment occurs according to the original combinatorial system, set by genetic laws.

Science, in D. Price's picturesque expression, is a thirteen-sexual system. Each new publication, on the average, is based on thirteen previous publications; these parent publications are referred to in a bibliography. Such a system of information interaction allows us to avoid some errors by eliciting logical contradictions—"contradictions of encounters," as they were called by Podgoretskii and Smorodinskii (1969). These create new concepts comparing the statements of previous, i.e., parent, publications. However, in contrast to biological systems, in science not all parent publications carry the same information load.

In both systems, the exchange of information has a fecundating character. In fact, it is the mechanism of fecundating information interaction which creates a self-organizing information system.

Communities in the Biosphere and in Science

Both in the biosphere and in science, the exchange of information occurs mostly within the framework of fixed, clearly outlined groups. In evolutionary biology the main idea is that of biological species (Timofeev and Resovskii, 1969). The most important characteristic is the complete biological isolation, i.e., uncrossability with other species. The creation of conditions favorable for biological isolation is one of the factors of species formation. Here, of course, we say nothing of such intricate mechanisms as the "horizontal" information transmission by virus transduction of genes, plasmid mechanisms, and so on, since their role in the process of evolution is not yet quite clear. Within the species, one can distinguish separate populations—groups of sexual organisms which occupy a certain area inside which information exchange, i.e., free crossing and mixing, is performed (Schmalgauzen, 1964). In a similar way in science, the elaboration of new concepts and, in connection with this, the intensive exchange of information take place first of all inside scientific schools or, recently, "invisible colleges." True, science is unlike the biosphere in that intergroup exchange of information is not forbidden but only hampered. The tendency to localize information exchange within specifically arranged communities is of the same kind both in science and in the biosphere. Thus, the biosphere preserves a successfully found idea which, however, has not yet been widely accepted.

Both in science and in the biosphere, we can also see the emergence of other communities, where there is mutual help but no internal exchange of information. In biology this is the well-known phenomenon of symbiosis, and in science it is the cooperation of basically different scientific communities in an effort to solve a single complex problem. In such a super-large community, each group solves its own specific problem and gives but limited results for common usage. For example, an analytical chemist transmits to metallurgists the results of his analysis, but he does not discuss with them the problems of analytical chemistry. Here, as in the case of biological symbiosis, the detailed exchange of information which provides the development of the communities in cooperation occurs only inside them.

The System's Freeing of Itself from Outdated Carriers of Information

Both systems have a "memory" which stores the essential phenomena acquired in the process of evolution. In science we can see these phenomena in textbooks; in biology, by studying the development of an embryo.

An information system should make itself free from ballast, that is, from out-dated information. In a biological system separate individuals—the carriers of the inherited information—die, and even the whole species may become extinct. In science it is the old publications—the primary carriers of information—that disappear. This disappearance of publications can be traced by their citation. If we draw a histogram of citations by marking the time which has passed since the publication of the paper on the abscissa and the number of citations on the ordinates, usually we observe an exponential decay in the number of citations. The half-life (which is similar to the period of semidecay in radioactivity) for various fields of knowledge usually fluctuates from five to about ten years.

In any case, at present in mathematics nobody cites the works of Newton and Leibniz despite the great significance they had for the development of mathematics, except, of course, when the topic at hand is the history of science.[7]

---

[7] In this connection it is noteworthy that the above-mentioned investigations of S. A. Zaremba demonstrated that in the early stages of the development of science (the eighteenth century), when it had not crystallized as a system, there were no definite signs of its self-purification at the expense of outdated publications. Roughly speaking, the old papers were cited as often as the new ones.

The System of Restrictions That Stabilize Development

An informationally developing system should possess a certain conservatism which would deter it from excessive variability; otherwise, it will lose its stability. How does the system regulating this process work? In what way is this conservatism balanced with the tendency for the development of new ideas or for the appearance of new forms? It does not seem that science is at present able to give any definitive answer to these questions. So far the necessary attention has not been paid to the study of the mechanism of this system of self-restriction.

Evidently, there is a very delicate mechanism which, to a great extent, sets the evolutionary process of the system and seems to be evolving gradually itself.

It is known that in a cell there is a mechanism for retreating from too strong genetic variations. If we regard the process of biological evolution in time, we shall find some points of rapid progress as well as some "dead-end" directions.

We know that there are about 8,500 species of birds and more than a million and a half different species of insects (see the preface by Huxley to Teilhard de Chardin, 1965).

At the same time, Man as a species does not form subspecies. There is a process of convergence which may be explained by migration or cross-marriages. Ramification has stopped here, but the process of development is still going on. The situation is quite different with the insects: extreme ramification and, apparently, cessation of further development are observed. Their "intellect" is fixed in simple reactions; their fragile scale has not allowed the appearance of larger forms. Too great specialization is one of the reasons for the appearance of the dead-end directions (Teilhard de Chardin, 1965). Of course, all of these statements do not always seem very convincing.

Now we shall see in what way the mechanism of self-restrictions operates in science. Along with the tendency to develop new ideas, there is a tendency to maintain stability. To a certain extent, scientists are always conservative, trying to preserve the existing concepts. A wonderful collection of historical examples have been selected illustrating the fact that the scientists' conservatism delayed for a long time the acknowledgment of a number of fundamental discoveries (Barber, 1961). In spite of the famous remark by Wiener (1964) that probably 95 percent of the original papers in mathematics are written by not more than 5 percent of scientists, one suspects that the major part of them would not have been written at all if the other 95 percent had not assisted in creating the high critical level. In science it is not enough to put out a new idea: one must over-

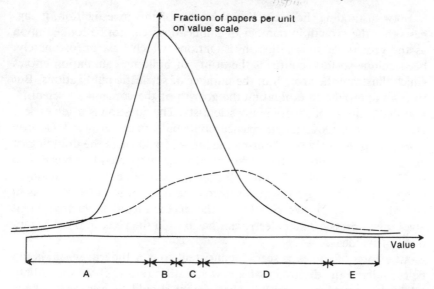

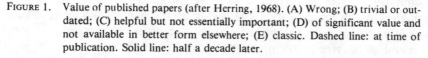

FIGURE 1. Value of published papers (after Herring, 1968). (A) Wrong; (B) trivial or out-dated; (C) helpful but not essentially important; (D) of significant value and not available in better form elsewhere; (E) classic. Dashed line: at time of publication. Solid line: half a decade later.

come the barrier of the intellectual field which prevents it from being accepted.[8] Although in the West there is no equivalent of the "guru–pupil" relation that has always existed in India (Parthasarathi, 1967), in science as in biological systems, a certain mechanism was elaborated for creating new ideas or new forms in the process of mutual fertilization. This mechanism is regulated by a certain system of barriers that protect the stability of this process.

We come next to a most important problem, but one that is difficult to pose as a question. Let us attempt to formulate it as follows: In what way does the content of science develop in time? Specialists in the measurement of scientific progress may answer: The number of publications is growing exponentially or, to be more precise, according to a sliding exponent, and publications are elementary carriers of information. However, the question of their scientific value cannot be avoided. C. Herring, a physicist, attempted to answer this question by analyzing publications in the field of solid-state physics. The results of his investigations can be seen in Fig. 1.

---

[8] Here we speak about the role which paradigms play in science. This question was considered in more detail in the first chapter of this book.

Notwithstanding the subjective character of this investigation, it suggests that the growth in number of publications cannot be looked upon as the growth of noise; this investigation—which has, unfortunately, been unique so far—confirms the utility of building cumulation curves which illustrate the growth in the number of scientific publications. But such an approach to evaluating the growth of the contents of scientific knowledge does not satisfy many scientists. The question is often raised: How do the discoveries in science contribute to its growth? Dobrov (1969) is apt to reply that "during the last 40 to 50 years the doubling of new scientific results in the world has been accompanied by an eight- to tenfold growth of scientific information. . ." Kompfner (1967) attempted to construct a histogram which showed the distribution of new ideas in electronics (Fig. 2). We clearly see the maximum in the histogram; it gives the impression that electronics has passed the peak in the development of its ideas.

All efforts of this type seem inconsistent to me. Indeed, what should be considered the discovery of a new scientific result? If we are able to establish a set of new results, what weight should be ascribed to each one? Why are equal weights ascribed to all discoveries and ideas in Fig. 2? If we consider such discoveries as the Raman effect or atomic fission, should we ascribe to them equal squares in a histogram or, as we say, equal weights? Can we consider them independent discoveries at all? We may suppose that one of them is a direct corollary of the ideas of quantum mechanics, and the other, of special relativity theory. One may try to evaluate discoveries by the number of later publications flowing from these discoveries. But in this case, we shall return to our starting point, namely, to the global counting of publications, because they are all caused by some new and important idea.

It seems to me that another formulation of the problem would be more appropriate: the study of the "dead end" or "barren" directions in science. These are the directions which in the course of their development do not create disputes because they do not lead to the uncovering of internal contradictions and to the creation of new concepts. The development of such barren branches of science can be easily observed in the spectrum of citation. If we build diagrams for a scientific discipline, putting the years that have passed since the moment of publication on the abscissa and "specific citations" (the number of citations per publication) on the ordinate, in such diagrams there would be no sharp peaks in barren branches of science.

Barren branches seem to appear in all spheres of knowledge. In mathematics, after an interesting theorem has been proved, there may be a lot of absolutely uninteresting works, narrowing, to some extent, the premises which were presupposed in the primary results. In some coun-

Period	Devices (milestones)
Before 1890	Geissler tube
1890–1900	Cathode-ray tube; X-ray tube
1900–1910	Geiger counter; Triode (audio); Thermionic diode rectifier
1910–1920	Multigrid tubes; Triode oscillator; Photocell
1920–1930	Thyratron; Electron diffraction tube; Magnetic electron lens; Iconoscope; Magnetron and Birkhausen-Murtz oscillator
1930–1940	Space-charge-electron gun; Klystron; Photo-multiplier; Electron microscope; Science of electron optics; Cyclotron; High-energy particle accelerator
1940–1950	Transistors and semiconductor device theory; Traveling wave tube and beam; Multicavity magnetron; Betatron, synchrotron, and linear accelerator
1950–1960	Masers and quantum electronics; Tunnel diode; Diode and electron-beam parametric amplifier; Strong focusing synchrotron; Ferrite nonreciprocal devices
Since 1960	Gunn-effect generator; Optical maser and ramifications

FIGURE 2.   Electronic device milestones.

tries, mathematicians, especially specialists in mathematical statistics, have begun to speak openly about the danger of the so-called "prestige" publications. These papers are written by the many well-trained mathematicians who lack creative imagination but feel that they must enhance their prestige. For this purpose problems are formulated with no concern for the utility of their solution or the reality of the initial premises. From the viewpoint of pure mathematics, these problems and their solutions prove of little value, though the authors make use of complicated and quite up-to-date mathematical methods. Similar situations are appearing in chemistry. The modern chemical industry has produced dozens of thousands of compounds; their number is growing exponentially in time. We may try to study the mechanism of the corresponding technological processes by studying the kinetics of reactions, the hydrodynamics of the attendant processes, etc. Actually, this is hardly necessary, for the technological processes can be optimized by statistical means (design of experiment) if there is insufficient knowledge of the mechanism of the phenomena. But it is tempting to carry out a thorough theoretical investigation. Work of this kind appears quite respectable and is greatly appreciated on the "exchange of scientific prestige." If such investigations are carried out on dozens of thousands of preparations, will it stimulate useful conflict? Another example is spectrochemical analysis, which has been known for more than a century. A lot of factors different in their physical nature are responsible for the development of this process. For many decades, scientists tried to project this set of factors into the space of the lesser number of variables and thus build a comprehensible general theory. It is clear that this problem, generally speaking, may have an infinite number of solutions. New publications appear in large quantities, but they do not in any way stimulate creative conflict.

Mul'chenko et al. (1979) made an interesting comparison of the informational activity in chemistry and physics (in the field of chemistry, the subject matter of the investigation was the activity of the leading scientists in its various branches; in physics, it was only those working in the field of low temperatures).

Chemists were characterized by very high productivity — from 106 to 755 papers during their working years. The average annual number of publications for a scientist varied from 3 to 19. For one of the scientists investigated, monthly productivity was more than a paper and a half. Such high productivity is achieved at the expense of what I call "ephemeral colleagues." These are graduates, postgraduates, and others who come within the leading scientist's sphere of influence for several years and then sink into oblivion. The response to such mass production proves rather odd. For the leading scientists–managers, the number of citations is very high: from 30 to 350 annually (according to *Science Cita-*

tion Index).[9] At the same time, the average citation of journal publications remains low; it varies from 0.4 to 1.87, and no single paper published creates an appreciable splash in citations.

The data were found to be quite different for physicists working in the low-temperature field. They were characterized by low productivity (from 1.7 to 3.7 publications per year), small numbers of publications with joint authors, and the absence of "ephemeral colleagues." Here again, high personal citation of the authors was observed (54 to 489 in journal publications), but this was accompanied by high specific citation of journal publications (1.86 to 13.33). In addition, the citation of some papers showed an explosive splash. Apparently, these papers were creating a revolutionary situation in their particular field.

Granovskii et al. (1974), in studying the informational activity of Soviet chemists–doctors of sciences, found informationally invisible scientists. For instance, in 1966, among 947 Ph.D.'s studied, 61 percent had no citations in native publications and 15 percent had only one citation each. It is true that in 1970 the number of Doctors of Science with no citations in the native literature fell to 30 percent (probably as a result of the increased number of Soviet journals scanned by *Science Citation Index*), but the situation still remains sad: a great number of scientists appear excluded or nearly excluded from the informational interactions. These data are presented in Table 1 in greater detail.

Shreider and Osipova (1969) suggested the term "intellectual industry" for works which create the branches of science with a barren character. This term is, to a certain extent, very appropriate: it emphasizes the fact that these publications are products of intellectual activity and undoubtedly have some intellectual value. Further, Shreider and Osipova suggest that perhaps in the future "intellectual industry" will develop in accordance with the economic needs of society. This assumption, however, must be stated very cautiously. In science there is no filter that would screen out the publications creating barren branches of science. Retrospectively, we may say that a given branch of science has so far remained barren, but this gives us no right to assert that this situation will be preserved in the future. Many examples can be given of situations when such forecasts, which had once seemed almost self-evident, were proved false by events. Under our very eyes, a branch of experimental design in mathematical statistics which for a long time had been looked upon as a mere "prestige" effort suddenly gave wonderful practical results (Nalimov, 1969). It is interesting to note that the system of filters

---

[9] Compare these figures with those from *Science Citation Index* for 1965. The average number of citations of a paper registered in the index was 1.65, and that of an author was 5.8. Nobel Prize winners (before winning) were cited 1.69 times on average. In contrast, leading scientists in the field of genetic code study were cited 112 times, and their co-authors were cited 42.5 times.

TABLE 1. *Distribution of Soviet Doctors of Science according to the number of times their work was cited (from Granovskii et al., 1974)*

| No. of citations | Percentage of Doctors of Science with the indicated no. of citations in: | | | | | |
| | Soviet publications | | | Foreign publications | | |
	1966	1968	1970	1966	1968	1970
0	61	51	30	44	37	35
1	15	14	12	17	17	17
2	7	9	7	9	10	10
3	5	6	7	6	6	7
4	3	3	6	4	5	5
5	2	2	5	4	4	4
X^*	1.6	2.8	7.0	3.3	3.6	4.4

* The mean of the number of citations per scientist, calculated for Doctors of Science who had no more than 20 citations. The number of Doctors of Science with higher values of this parameter did not exceed 5%. The tail part of the function is omitted from the table. There were Doctors of Science with numbers of citations of 30 or higher.

in science is arranged in such a way that it is easier to publish articles which create barren situations, i.e., which stimulate no conflict. These articles irritate nobody, they make defense of a thesis easier, they can be published in large quantities, and they help to enhance the writers' prestige.

Returning to the comparison of science and the biosphere, let us look again at the processes of development. A single species may grow exponentially for a time but later show a saturated curve. Separate members of this species are ordinary carriers of information, comparable to publications in science. We might ask what law governs the growth of the number of species and families in biology, but the answer to this question is not readily forthcoming. In biological evolution we speak about dead-end branches, and if we equate these to what I have called barren situations in science, the essential differences in the growth of science and of the biosphere immediately become obvious. In biological evolution there is a certain mechanism which takes control when the development of a species proceeds in unfruitful directions. Further development in these directions comes to a stop, and these branches seem to stiffen and turn into the dead-end of the evolutionary tree.

In the evolutionary process, many tentative steps are made, and the unsuccessful ones are abandoned; they never ramify indefinitely. I mentioned above that there are more than half a million insect species, but this phenomenon is unique; it has not become typical of the evolutionary

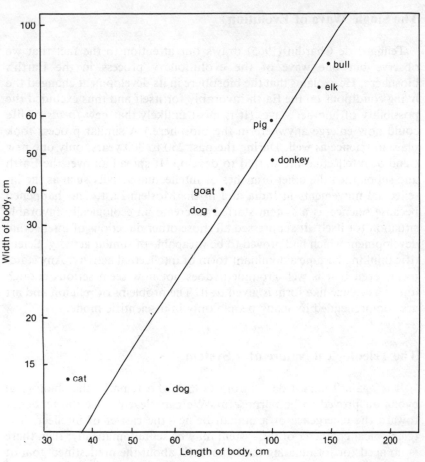

FIGURE 3.   Correlation between length and width of bodies of some members of the Mammalia (after N. Rashevsky).

tree, and we might ask why this is true. In biology there is a well-known mechanism of uneven (allometric) growth, studied thoroughly by J. Huxley (1932). It appears that the correlation of different parts of an organism, e.g., the correlation of length and width of the body of the Mammalia, is connected by the relation $y = ax$, as is illustrated graphically in Fig. 3. The law of allometric growth is connected with the optimality of biological systems (Rosen, 1967). It is important to pay attention to this phenomenon, as it is one of the manifestations of the mechanism which represses the development of too great a variety of forms.

The Single Wave of Evolution

Teilhard de Chardin (1965) draws our attention to the fact that we observe but one wave of the evolutionary process in the Earth's biosphere. He explains that the biosphere in its development changed the living conditions on the Earth favorably for itself and thus excluded the possibility of another wave. (It is most unlikely that new *forms* of life could now emerge anywhere in the biosphere.) A similar process took place in science as well. During the past 250 to 300 years, only one new trend of intellectual life started to develop. It spread all over the Earth and supplanted the other branches of intellectual activity such as the intellectual movements in India and in the Moslem East. This happened because science as a system started to create an ecologically favorable situation for itself. It suppressed all those other directions of intellectual development which had proved to be incapable of similar activity. Scientific thinking became a dominant form of intellectual activity. Any statement, even if it is well grounded, does not now seem serious enough unless a science-like form is given to it. The problems of religion and art are comprehended by many people only in a scientific mode.

The Teleological Nature of a System

It is again Teilhard de Chardin (1965) who remarks that biological evolution proved to be purposeful. We can clearly trace the tendency toward the emergence of a human being—the carrier of intellect. The teleological character of the system may be set automatically; then there is no need for formulating the hypothesis about the predestined goal of the system.[10] This view about the goal of evolution raises sharp objections from many biologists. Perhaps we should word our statement more carefully: the increasing complexity of the intellectual structure of organisms may be considered an evolutionary stride forward, and this process, clearly traced in time, leads to the appearance of Man, who influences more and more the rebuilding of the Earth's biosphere. In science we can also trace (at least retrospectively) the goal of its development, which manifests itself in the development of ever more complex hypotheses and theories.

---

[10] At any rate, speaking of self-organizing systems, we should always subject to analysis the finite goal of the system, ascribing to it some conventional meaning. We need the concept "goal" to build a logically complete model of a system.

A Glance at the Future

It is impossible to predict the future of informationally developing systems. This follows from Gödel's theorem on the undecidability of logical systems. A sufficiently strong informationally developing system will contain those truths which cannot be derived from the initial premises by means of the finite number of deterministic derivation rules. Consequently, the only thing we can do is to discuss what must or, more precisely, may happen to a system if there is no change in the organization of its control, which we cannot forecast.

A system may develop as long as there is sufficient nutrient medium for it or, to be more exact, until it is capable of changing the environment in the necessary direction with sufficient speed. It appears that now, for the first time, the Earth's biosphere faces serious ecological difficulties. This has occurred only after the climax of its creation — Man — started to develop science. Little by little, interaction between science and society, as well as between society and nature, became dangerous. Many serious papers are now written about the coming exhaustion of the globe's natural resources (see, e.g., Watt, 1968).

As to the resources for science, the state of affairs is much more serious. I already mentioned that the expenditures on science grow more rapidly than the income of society. It is interesting that the rapid growth of expenditures on science seems to be to a great extent connected with the fact that science in its development could not work out a filtering system to screen out barren branches. It is probably impossible to estimate quantitatively the debilitating influence of these branches of science upon the organism of science.

It is important to note that it is here that the difference between the two self-organizing systems is extremely acute. In the biosphere, unlike in science, the system of restrictions appeared to be constructed in such a way that the dead-end directions are screened out before the external nutrient medium has been exhausted.

Will filters of this kind appear in science in the future? This question cannot, of course, be answered with confidence. Computers may turn into such filters. Perhaps, under the influence of computers, a reappraisal of values will take place in science. Introduction of computers directly in the sphere of scientists' intellectual activity will at last permit us to draw a strict borderline between a routine and a creative component in a scientific worker's activity. Even now it is possible to use computers for solving an intellectual problem which seems very complicated, that is, for choosing the optimal method for complex organic syntheses (Corey and Wipke, 1969). Might not the work of a specialist in organic chemistry turn into purely technical, almost routine activity? A great deal of experience has been accumulated in the use of computers for solving mathematical problems which previously were considered purely intellectual. In the field of engineering research, such branches as the design of extremal experiments and the estimation of their results may at least partly be handed over to consulting computers. If everybody understands that the intellectual industry, or at least part of it, is connected with purely routine activity, scientific workers—and first of all the younger ones—will lose interest in it. Spontaneous and uncontrolled influx of free forces in those fields of science will cease, and they will develop only to the extent of the real needs of society. This is the way all branches of engineering are developing now.

Who knows whether this will really happen? We can only discuss the possibility of such a process. And here again, we must emphasize the great difference in the development of the biosphere and of science. In the biosphere the restricting mechanism is in some way connected with natural selection, that is, with external information. In science we may expect the emergence of the restrictions that will restrain its development only after the reappraisal of values has taken place. Here we should take into account the psychological conditionality of the system.

In this chapter I have applied the cybernetic approach to the study of complicated systems. An effort was made to understand the functioning of systems by means of studying the systems of control located within them, i.e., the structure of the mechanism which makes these systems self-organized. Naturally, we had to disengage ourselves from many very essential properties of the systems under consideration. Whether such a high degree of abstraction is reasonable, whether it has heuristic value, whether it gives an impetus to really interesting research, only the future will show. At present, we are taking the first timid steps in this direction, and I would like the reader to regard this publication only as one of these steps.

Chapter 8

The Problem of Complexity in Describing the World Scientifically[1]

A Formal Analysis of the Difficulties in Constructing Theoretical Biology

Psychological Grounds for Judging Whether a Statement Is True

Great efforts have been directed at developing the concept of the *logical truth* of statements. The approaches of Tarski and Carnap are well known (see, e.g., Gastev and Finn, 1964). However, in science, as in our everyday life, when we acknowledge the legitimacy of judgments, we are prone to proceed not so much from their logical truth as from premises of a psychological nature.

In my experience, a statement acquires the right to be called scientific when at least one of the following conditions is fulfilled.

Condition 1. A compact set of concepts is formed. This allows the making of judgments of the phenomenon under study in a compact form. In the natural sciences, we then speak of having created a theory. Theory, in some formal meaning of word, is a type of logical structure which allows us to describe the phenomenon observed in an essentially briefer manner than could be done without any theoretical considerations, after immediate observation. Even a compact presentation that does not contain any theoretical considerations is regarded as a type of theory. Mendeleev's periodic table of the elements, at the moment of its

[1] A brief version of this chapter was published in the abstracts of the VIIth All-Union Symposium on Logic and Methods of Science, Kiev, 1976, pp. 234–236. This chapter was translated by A. V. Yarkho.

An Attempt to Grasp the Reality with a Set of Models

appearance, did not contain any theoretical reasoning, yet it was part of chemical theory. The same is true of Linnaeus's classification.

In mathematics, a compact presentation of judgments is embodied in mathematical structures. According to Bourbaki (1950), mathematics differs from other sciences in that judgments made within it can be reduced to mathematical structures—compact constructions rich with logical consequences. Strictly speaking, we should call mathematical science not all groups of statements written in mathematical language but only those which display mathematical structures. For example, the theory of probability acquired the status of a mathematical subject only after Kolmogorov gave its axiomatic structure. If we hold to the concepts developed by Bourbaki, we should designate as mathematical models of the external world not every description expressed in mathematical symbols but only those with rich structural content. Theoretical physics, whose structure I shall discuss in detail below, has proved to be arranged so that its content is given by fairly compact statements made in the language of mathematics. By the way, here lies the reason for the oft-repeated statement that the structures of physics are the model for all other sciences.

The construction of compact statements in the form of mathematical models seems possible in other branches of knowledge, including the humanities. As an illustration, I mention my study of language semantics. I suggested that the mechanism of speech comprehension be described by means of the Bayesian theorem. Later, I myself was amazed by

The Struggle Against Complexity

the great theoretical purport of this approach, which allowed various peculiarities of verbal behavior to be easily explained. Moreover, certain concepts of the nature of thinking itself arose easily (see Nalimov, 1974*b*, 1979, 1981). At the same time, expanded mathematization of knowledge in some cases resulted in the construction of the so-called "portrait models," which do not have any new content but simply express in mathematical language what can just as well be expressed in common language. It is only too clear why such models annoy representatives of concrete branches of knowledge: in these cases, one language is unjustifiably substituted for another one. As an example, one can ask how biology was benefited when some of its ideas were reformulated in terms of information theory.

Nonetheless, compact presentation makes possible many rich consequences. If a complicated phenomenon is expressed by a compact model, we always have a basis (psychological rather than strictly logical) for believing that such a description might include things which we have not yet observed but will be able to observe by making special efforts. In this way theories acquire their predictive power — the possibility that they may predict yet unobserved phenomena. Here lies the principal difference between modern science and alchemy. The latter described its phenomena in a mytho–poetical language, the way they were observed, and for this reason it could not predict new phenomena. The progress of European culture is largely due to linguistics, since therein was developed a symbolic language of compact ideas.

Compact presentation allows us to simulate all possible evolutions of a phenomenon. In this case it will suffice to vary the initial conditions. More than that, compact presentation allows us to control the phenomenon, which always gives the illusion of complete comprehension, i.e., of cognition. The possibility of controlling something as it is cannot be acknowledged to be the criterion of truth. Man had learned to control some technological processes (e.g., metallurgical ones) long before any scientific concepts of metallurgy appeared. A more detailed critical analysis of the concept "cognition" was given in Chapter 1 of this book. There I did not give an optimistic answer to the question of what cognition is. Here I propose the following answer: formally speaking, cognition of the world is a possibility of recording in a compact form the observed phenomena. But this answer will hardly satisfy everyone. Again, this would mean that cognition is made equal to mastery of the world, since compact presentation is just what allows us to forecast and control.

Condition 2.   Past knowledge is being strengthened. In Chapter 2, I spoke of our knowledge always being probabilistically weighted: we are apt to ascribe greater probability to some judgments and smaller probability to others. New assertions obtained as a result of the analysis of newly observed phenomena acquire the features of scientific knowledge only if they strengthen the previously existing theoretical structures. The mechanism of placing new statements in correspondence with old knowledge may be described by a probabilistic model built in the system of neobayesian notions, of which I already spoke in Chapter 2.

The theorem of Bayes,[2] well known in mathematical statistics, provides us with a mathematical model of the way previously accumulated knowledge (probabilistically weighted) mixes in our mind with that newly acquired.

The development of scientific schools can be understood easily from this standpoint. First, the leader creates a fuzzy and insufficiently confident concept accompanied by doubts and reservations. In this fuzzy field of elementary events, the probability ascribed to separate judgments is determined in some way. The leader's pupils try to get new knowledge that will narrow the initial set of fuzzy concepts. At a certain moment, all the probability proves to be concentrated in a very narrow subset of the set of initial statements. The distribution becomes almost needle-shaped. Further strengthening of the correctness of the narrowly concentrated statements becomes unnecessary, and the school becomes inactive. And if we look at the publications of the members of the school, we see that their articles always end with the following incantations: our results "are in good agreement with . . . ," "support previously formulated hypotheses . . . ," "they do not contradict . . . ," "strengthen the previously obtained results." Everyone knows how pleasant it is to refer to the fact that the idea you are developing has already been mentioned somewhere, though its formulation was then rather weak. Within some paradigms, this form of presenting ideas is the only one possible, since only in this way do they acquire a scientific character.

Condition 3.   Unexpected new statements evoke wonder. Wonder is a term used by Plato and Aristotle. In his *Theaetetus* Plato (1953) wrote, ". . . for wonder is the feeling of a philosopher, and philosophy begins in wonder." Aristotle's *Metaphysics* (1966) includes the words, ". . . for it is because of wondering that man began to philosophize and does so now." Bohr uses the expression "a mad hypothesis." The possibility of explaining the old in a new way, of observing it with new eyes, immensely attracts some people and arouses opposition in others. Its attractiveness lies in the fact that the new, unexpected view of things is a new guessing which might help us to see what has been concealed from us by the old system of concepts. Opposition is caused by the fact that a novel view is

---

[2] In my earlier books (Nalimov, 1974*b* 1981) I used it to explain the mechanism of comprehending phrases built over words with fuzzy prior semantics.

always a heresy—the refutation of what has been recognized as scientifically true.

The two above-mentioned positions are mutually exclusive, and their supporters are people with different genotypes. (On behavioral genetics, see, for example, Davis, 1975.) The struggle of opinions in science is very often only a conflict between people with different psychological characteristics.

If conditions 2 and 3 presented above are characteristics not only of scientific activity but also of all other human intellectual activities, the first condition—compact presentation—is an exclusive prerogative of scientific activity. European culture is primarily an abstract–symbolic, i.e., compact, record of our knowledge of the world.

Compactness of Theoretical Constructions in Physics[3]

In an earlier book (Nalimov, 1981), I presented the results of a study designed to determine the degree of saturation of physics texts by mathematical symbols. This I referred to as the "symbol complexity" of publications in physics. For my purposes, "words" were represented by mathematical operations such as the derivative dy/dt, the inverse matrix $\mathbf{M}^{-1}$, etc., and "phrases" consisted of symbols and the operations performed with them that are separated from the rest of the text. For example, a phrase is a mathematical expression such as

$$m = \sum_{i=1}^{n} a_i m_i$$

The study was carried out from an historical perspective. It included six books on general physics published in 1760, 1797, 1833, 1874, 1933, and 1948. In addition, four books on field theory and quantum mechanics published in 1931, 1946, 1948, and 1949 were considered. By calculating the average number of "words" and "phrases," as defined above, on a standard page of text, I was able to show that the symbol complexity in books on general physics remained more or less constant from the beginning of the nineteenth century through the rest of the period considered. No increasing trend toward saturation of the texts with symbols was observed. For publications in field theory and quantum mechanics, the situation was slightly different in that the texts were

[3]This section and the next one were published in 1977 in the Polish journal *Zagadniena Naukoznawstwa* and in Ukranian in the journal *Avtomatika*. The translation used here is borrowed from the journal *Industrial Laboratory*, which is regularly translated in the United States.

richer in mathematical symbols, although the difference between the individual books was small.

These results led me to conclude that differences in the complexity of the contents of books do not manifest themselves by a difference in symbol complexity. Although the language of physics is undergoing a steady evolution, the text becomes more complicated not as a result of an increase in the number of mathematical symbols or phrases but as a consequence of the greatly expanded meaning carried by these symbols and phrases. Thus, the same symbol can denote a scalar, a vector, a matrix, or some other quantity. The development of physics, i.e., the expansion of its scope and meaning, takes place in such a way that the compactness of its symbols is preserved.

Let us write down some well-known formulas of modern physics.

The Lorentz transformation:

$$x'_1 = \frac{x_1 - vt}{\sqrt{1 - v^2/c^2}} \; ; \quad t' = \frac{t - x_1 v/c^2}{\sqrt{1 - v^2/c^2}} \; ; \quad x'_2 = x_2; \quad x'_3 = x_3$$

The Schrödinger equation:

$$i\hbar \frac{\partial \psi}{\partial t} = \mathbf{H}\psi$$

Heisenberg's uncertainty principle:

$$\Delta p \Delta x \geqslant \hbar/2$$

These very short formulas carry a very large amount of information and have given a considerable impetus to the development of modern physics; in particular, the first of these formulas expresses the theoretical feasibility of an atomic bomb, with all its consequences.

The search for compact formulas has prompted physicists to use probabilistic considerations in their theoretical constructions. In deterministic systems it is not possible to describe by a compact formula the behavior of molecules that are in a gaseous state.

In contrast to the opinion held by the philosophers of the past, the necessity of describing phenomena in probabilistic language was not a result of lack of knowledge but rather of the need to express knowledge in a compact form. Here it seems appropriate to recall the algorithmic definition of randomness. A finite sequence of numbers is called a random sequence if it is not possible to construct for it a generating algorithm that could be written in a form shorter than the sequence itself. [The algorithmic method of definition of randomness is considered in more detail by Fine (1973), who also presents a detailed analysis of the logical foundations of all (or almost all) the probability theories available nowadays. Here I deliberately present a somewhat simpler formulation, by proceeding from a theory not yet fully developed.]

Precisely such a situation was encountered in the construction of statistical physics and thermodynamics. The statistical approach made it possible to describe the behavior of a system with the aid of average quantities, without describing the behavior of each individual molecule.

Is a Compact Description of Knowledge Possible in Biology?

From a formal point of view, the difficulties encountered during the entire period of the development of biology are due to the need for a compact description of the huge amount of material readily accumulated as a result of observations. The first successful attempt to describe the great diversity of observations was the classification of Linnaeus, which, however, was extensively revised later on. Darwin's evolutionary theory is an attempt at a compact classification of these same data, but from an historical perspective. However, from Darwin's time to our own, we have not found any all-encompassing and compact theoretical constructions with the same scope of interpretation as, for example, the formulas of theoretical physics presented above. The discovery of the biogenetic code (the most important discovery in biology in recent times) is in fact the deciphering of a language, not an explanation of how something *new* is written in this language. Here the success achieved consists in finding in biology a structure familiar to our intellect, i.e., a language, and analyzing it formally.

Many publications have appeared recently in which mathematical models of biological phenomena are constructed. There even exist special journals devoted to mathematical methods in biology, for example, *Biometrics* and the *Journal of Mathematical Biology*. It would seem that methods have been found that lead toward a compact representation of knowledge in biology. But in fact matters do not work out as simply as that. In any case, the publication of an ever-increasing number of papers on biological subjects that contain mathematical models does not bring the structure of biological science any closer to the structure of physics. What are the reasons for this?

Mathematical models in biology, just as in many other branches of science such as psychology and sociology (Harran, 1963; Szaniawski,

1975), can be divided into two classes: descriptive and theoretical (in psychology and sociology the latter are sometimes called prescriptive models). In the first case, we have models for accumulating and compactly representing the experimental data. Such descriptive models are constructed without penetrating into the essence of the phenomena under study. In the case of observational results expressed by a matrix, it is possible, for example, to represent these data with the aid of the principal components, and if it turns out that the major part of the total variance is contained in the first components, we thus obtain a considerable compression of the observational data. It may happen that the first components can be easily interpreted theoretically, but this may also not be the case, and then we adopt the method of principal components as a formal procedure for reducing the amount of data to be studied. Even a simple calculation of the sample mean and of the confidence limits for these data requires a (possibly very simple) mathematical model. A long time ago, R. Fisher defined mathematical statistics as the science of reduction of data. Note once again that the reduction of data by statistical methods can take place without penetrating into the essence of the phenomena underlying these data. This is both the strength and the weakness of statistical methods. Statistical models of this type do not yield a compact representation of our knowledge about the phenomenon under study, but are merely a compressed form of the observational results. Such a compressed representation of data naturally sharpens the intuition of the investigator.

Theoretical models claim to reveal the "mechanism" of a given phenomenon. They are constructed as deductive structures based on clearly formulated premises. They are said to be prescriptive, since they prescribe a norm of behavior that follows from these premises. In psychology and sociology these models are not so much verified as they are compared with the actually observed forms of behavior, so as to ascertain the extent to which it differs from a strictly rational (in the words of Carnap) behavior in a fully determined situation (Szaniawski, 1975). In the vast majority of cases in biology, models of this type are never compared with actual phenomena. None of the articles devoted to mathematical simulation in biology and medicine published in *Voprosy Kibernetiki* (Problemy biomeditsinskoi kibernetiki, 1975) contains any hint concerning the verifying of models by comparing them with actually observed phenomena. Many mathematicians are inclined to believe that such verification is superfluous. This also has its good reason, because if we try to verify such a model by comparing it with observations, this will be entirely ineffective. This is due to the fact that, in verifying these models, we cannot subject them to a crucial experiment. The observed phenomena are usually structured in such a way that their actual mechanism cannot be revealed. Thus, in studying the mechanism of a chemical reaction, it is very easy for the investigator to observe the time variation of the end product of a reaction, but it is very difficult to follow all the intermediate reactions. Matters are even more complicated in psychology, sociology, and (usually) biology. It almost always turns out that the experimental data can be adequately (in the statistical sense) described also by much simpler models. What, then, is the purpose of prescriptive models in biology?

The answer to this question is very simple. It is of interest, and often also very meaningful, to examine how a system would behave if its mechanism of behavior were as assumed by the designer of the model on the basis of premises that are to a certain extent probable, but nevertheless arbitrarily selected. Then it makes sense to compare, at least at the verbal level, this model of behavior with the considerations of the biologists dealing with this problem. Such a dialogue can be of interest. But even if it does not take place, the mathematician would still be interested in finding out what can follow from such (often very simple) premises. Indeed, sometimes the results are very interesting. For example, the periodicity of development of a population, well known to biologists, is a simple consequence of the properties of the model in the "predator–prey" problem.

Mathematical models of the first type can be understood on the basis of a fairly small amount of mathematical knowledge, and hence can be fairly easily grasped by biologists. In any case, the use of statistical methods in biological research is steadily increasing, and biometrics is a mandatory subject in biology departments. Matters are entirely different for models of the second type: most biologists are simply incapable of understanding them, since this would require a mathematical education at least equal to that possessed by physicists. At present we could not imagine physics, even experimental physics, without a mathematical foundation that would make it possible to understand the ideas of electromagnetic field theory, quantum mechanics, and relativity theory. But neither biologists, nor mathematicians dealing with the construction of mathematical models in biology, feel obliged to insist on a major extension of the teaching of mathematics to biologists. I. G. Petrovskii, one of the authors of the well-known paper "A Study of the Equations of Diffusion Accompanied by an Increase in the Amount of Matter, and Its Application to a Biological Problem," published in 1937, served for many years as rector to Moscow State University, but he did not try to raise the level of mathematics taught to biologists to equal that taught to physicists. The mathematics department has a course called mathematical biology, but no corresponding course exists in the biology department.

This has a simple explanation. Mathematical models of the second type, which claim to describe the mechanism of phenomena, do not have such a wide scope as the mathematical models of physics. The logical applicability of such models is not sufficient to justify the intellectual effort needed for a serious study of mathematics.

A compact description of biological systems that would encompass the entire complexity of their behavior is impossible, since [as was pointed out very convincingly by Monod (1972)] the world of biological phenomena can and must be described not in terms of necessity but in terms of chance [from a slightly different point of view, we have also written about this (Nalimov and Mul'chenko, 1970)]. In other words, the complexity of biological phenomena is such that it cannot be described more concisely than by writing down all the observed phenomena. On the basis of the results of a short series of observations, we cannot write an algorithm that would express (even approximately) the subsequent evolution of the system.

Let us examine this assertion in detail. One of the peculiar features of biological systems consists in the possibility of observing such phenomena at two levels, as it were. One of them is the surface level, when the phenomena take place under certain steady-state external conditions; the second level is the deep gene–molecular level which manifests itself when the conditions of existence of the system change sharply. The phenomena taking place at the surface level are to a certain extent amenable to a brief description, but the knowledge about them is not of particular interest. Let me illustrate this with an example.

Suppose that we are studying a body of standing water, for example, a lake or even an ocean. The phenomena taking place in such a body while it is quiescent could no doubt be described in some compressed form by a system of differential equations (the parameters of these equations being the rates at which certain species eat up other species). It is true, though, that such models cannot be subjected to a crucial experiment when comparing them with the results of observations, and therefore the justification of such models is doubtful to many. But this has already been pointed out above. Here, let us note something else: if the conditions of existence of the body of water change, then a model describing the phenomena at the surface level will be useless. If sharp changes in the geological or meteorological conditions occur, or if a large amount of impurities (for example, oil) is poured into the water, then a random generator begins to act at the gene–molecular level that could not have caused major disturbances in the steady state. Processes take place at the molecular level that are due to structures of an opposite character, namely, mutations — expressions of randomness on which the rigid and unambiguous grammar of the genetic code is superimposed. Here we witness a profound analogy with the language behavior of humans, where the randomness manifests itself in the fuzziness of the meaning of words; the grammar superimposed on this fuzziness is one of the forms of expression of Aristotelian logic. Note that a mathematical description of phenomena in biology is most effective in the description of the heredity laws, since in this case the object of the description is a strictly ordered grammar.

We can say that the nature of change in biology is random, since it is impossible to find an expression for a sufficiently detailed description that is considerably shorter than the "most complete" description of the observed phenomenon. In other words, it is not possible to construct a model of a generator of mutations in terms of ordinary cause–effect relations, i.e., it is not possible to find the causes that unambiguously generate the full diversity of observed mutations. Having found that the nature of change is random, we are greatly surprised that there does not exist an ordinary probabilistic description of the observed phenomena. An ordinary statistical description of phenomena is possible if, on the basis of the results of observations carried out on a small sample, we can calculate the distribution parameters which make it possible to obtain an idea of the behavior of the complete sequence of phenomena. In the case of biological changes, observations made on a small sequence of phenomena do not yield information about the subsequent behavior of the system. In such a case, averaged characteristics have no significance. The individual manifestations of the phenomena are important, irrespective of their probability of occurrence.

Since the processes take place over a prolonged period, and since they encompass a large number of biological entities (carriers of mutations), it can also happen that events of very small probability are realized. These events can have very great and entirely unforeseen consequences if, as a result of mutations, we obtain features that are adapted to new conditions. We have such a situation when bacteria appear in a stretch of sea polluted by oil that are capable of decomposing the oil, thus poisoning the water with the decomposition products. Another such situation is the appearance of microorganisms that are capable of "eating" antibiotics; there recently have appeared strains of organisms whose normal development requires streptomycin. This is an example of the many unpleasant surprises with antibiotics. The response of microorganisms to changing conditions is amazing, and it virtually takes place under our eyes. Bacteria have become resistant simultaneously to four medical drugs, namely, streptomycin, chloramphenicol, tetracycline, and sulfanilamide. The resistance of bacteria is accounted for by the so-called R-factor, which can instantaneously propagate over the entire population; it is not specific, and this is especially worrisome. The possible consequences of the danger arising from this are difficult to estimate (for more details see, for example, Bogen 1967). In research (reported by my colleagues) carried out with patients suffering from chronic tuberculosis, phlegm cultures were treated with many drugs according to the rules of experimental design. The result was surprising, in that the combinations of drugs found to be critical were those that made no sense from a medical point of view. It is as though the microbes have guessed the thoughts of the doctors and do not perform as they are expected to do.

The difficulty in reducing our knowledge of biology to a compact form can be formalized as follows. On a field of elementary events we are given some events with a very small probability that are essential by their consequences. If one of them is realized, there appears another field of elementary events with another probability distribution. It hence follows that certain low-probability events may trigger other events with a high probability. For example, doctors assert nowadays that the appearance of even a single malignant cell is sufficient for the development of cancer. The probability of the occurrence of a single such cell in a certain organ of a person in a certain time interval is perhaps small. But once such a cell has appeared, the field of elementary events will change very rapidly, and with a high probability we can expect very well-defined ill effects. In the case of ontogeny we possess a large number of observations from repeated phenomena; therefore, we can make a probabilistic estimate of what will happen after the realization of some low-probability situation. In the case of phylogeny we do not possess any information about what will happen in a new situation realized as a result of some low-probability mutation. This can be reformulated in terms of conditional probabilities, but it will not facilitate our task. Monod (1972) has pointed out that, before the appearance of life on earth, its a priori probability of occurrence must have been equal to zero. Teilhard de Chardin (1965) attaches great importance to factors that would indicate that evolution develops toward less and less probable structures.

At present, rumblings can be heard that the existing language of mathematics is insufficient for describing biological phenomena, and it is necessary to elaborate a new (entirely separate) branch of mathematics especially suited to the simulation of biological problems. In my opinion, it is not the language that is at fault; rather, we have here a situation in which the past does not give us any information about the future. The complexity of the system is maximal, and in this sense it is a random system.

What can we say about the randomness generator that accounts for biological evolution? Where is it materially incorporated? What is its mathematical model? Is it equivalent to a one-dimensional sequence of numbers with a spectrum of the white-noise type, or is it something more complicated? We know all the difficulties encountered in designing a random number generator for a computer; in this case we try to conceal our ineptitude by pointing out that we are dealing only with pseudorandom numbers. Anyone who simulates problems on a computer knows that special attention should not be paid to "fine" effects, i.e., effects that may be caused by a violation of the randomness in a sequence of random numbers. But how are fine effects accounted for in biology, by pure randomness or by a violation of this randomness? In this way we start an inadmissible play on words, since we know too little about the nature of chance to be able to go into details. If randomness is interpreted as maximum complexity, then all this discussion becomes meaningless.

For a long period of time, beginning with Aristotle and continuing perhaps until the end of the nineteenth century, the philosophers and many scientists were of the opinion that our desire to describe something in terms of chance is due to our ignorance. But at present, when we define chance as maximum complexity, is this not simply a reformulation of our previous assertion? It seems to me that this is nevertheless something much more essential, namely, a change in our paradigm—the acknowledgment that the impossibility of describing something is due not to our ignorance but to a complexity which does not lend itself to description in principle. Take a simple illustration. If we are going to transmit the text of the Soviet newspaper *Pravda,* we have to do it word by word, without omitting anything. The reduction of the text is not possible though its contents seem perfectly clear and familiar. Perhaps some will say that we have not advanced very far, since we still do not understand the essence of what we cannot describe but have merely found for it much more important reasons than a simple acknowledgment of our ignorance.

In any case, all the talk about the possibility of associating biological change with hard radiation and other such factors does not in fact explain anything. We are simply dealing with certain triggers that either turn on or speed up the operation of a random generator unknown to us.

For a better illustration of this analysis, consider an example of a complex structure generated by a random generator about which we know something. The well-known caves in New Afon of the Caucasus, which constitute an entire sequence of dwellings, appear to express a unity of thought, an inner harmony. Some of the caves resemble temples, and other are like antechambers of temples. One senses a frozen rhythm cut in stone. Here are also curtains that have stopped moving, like petrified streams that are fancifully entwined; then we can see stalactites and elusive beings on the walls, as on those of Notre Dame.

> The caves . . .
> They are sudden,

They are beyond expression . . .
Ah! measured now by the height of the arches,
now by long passages,
now by whimsical forms,
by their perfection
which is so fresh, complete, and mysterious,
so finely felt
that only breathing broken
when it comes deep from the bottom of your heart
and aspires aloft
might express the sensation
which does not yield to words,
an inseparable welding of ecstasy and tremor
pronounced by the soul
which has submitted to monumental
and sublime silence,
so tangible and solemn
that you begin to feel the supreme work
going on in the caves
under the cover of silence.
You begin to hear its rhythm taking shape of a hymn
and you start to make out the words
"Let the flesh keep silence" (so that it might become
the witness to the miracle).

I shall attempt no further description of these caves: words cannot interlace so whimsically and marvelously, so silently and solemnly.

These caves were generated by two different factors: on the one hand by the physicochemical processess of dissolution and crystallization which are amenable to a rigorous description in terms of cause–effect relations, and on the other hand by randomness due to the inhomogeneity of the mountain rocks, their spontaneous movement, and the long-term seasonal changes in meteorological conditions. We could say that the language of physicochemical relations is used for reading the information from a random generator which represents the behavior of the lithosphere. The result of this reading manifests itself in the text, i.e., the caves. Here we know something about the location of the random generator, though it is difficult for us to imagine its mathematical model. In any case it is obvious that it must also contain fragments of periodic components related to meteorological processes. With regard to randomness, I shall note here only its property of complexity; i.e., neither the caves nor the factors generating them can be attributed to some simple process. Our description of them could not be much simpler than the possible description of the entire manifold of phenomena. The foregoing could be interpreted also as follows: We have in front of us an artistic architecture that came about by the play of chance and, more importantly, which appeared spontaneously, without sifting of the best versions from the set of randomly generated versions, as in biological evolution. The architectonic unity prompts one to think of teleology, but from the point of view of common sense this thought is quite absurd.

Now let us return to the biosphere. What could we actually say in this connection with regard to a generator of randomness? Is this simply the introduction of errors according to the laws of roulette into texts previously written in the language of the genetic code, or does it perhaps consist in a discrete reading of texts from information flows whose complexity is such that they could be described only in terms of randomness? If the entire evolution consists only in the introduction of errors into a previously written text, then this text must have been very complicated at the moment of its inception. In an earlier paper (Nalimov, 1979), I put forward the hypothesis that our intellectual activity takes place at two levels, namely, at a discrete–logical level of language, about which we know a lot, and at a continuous (extralogical) level which we could imagine as a continuous stream of consciousness. There exists a direct contact with this continuous stream during sleep, in certain hypnotic states, during creative activity, and in religious meditation. Our everyday language behavior is constructed in such a way that, in my opinion, the interpretation of our utterances takes place at the extralogical continuous level. All this is set forth in detail in the paper mentioned (Nalimov, 1979). Here I merely intend to draw attention to a possible analogy: If the creative process of human thought consists in discrete reading from a continual flow, then could it not be that biological evolution which leads to the appearance of new "texts" in the biosphere is a discrete reading from this same flow? If this is the case, then we have a profound analogy between the process of creative thought and the development of the biosphere.[4] Our idea that chance is maximum complexity also admits such an interpretation of the nature of a randomness generator. All this, of course, could be dismissed as belonging to the realm of fantasy, but nowadays it is customary in the philosophy of science to claim the right to put forward hypotheses as fantasies (free assumptions) that can subsequently be discarded if found unsuitable.

In any case, we have now reached the stage when it is necessary to try to formulate some (even very hypothetical) ideas about the nature of a randomness generator in biology. Human activity was found to be directed toward the creation of conditions favorable for the starting of a randomness generator. Such conditions appear, on the one hand, as a

---

[4] This is in agreement with the ideas of Teilhard de Chardin (1965) concerning the noosphere (the thinking layer of the universe) and evolution as an arrow of biogenesis directed toward the highest point "omega" that is the final accomplishment of everything. The contradiction between the probabilistic nature of mutation and Berg's nomogenesis (1969), assumed in the literature, disappears.

result of changes in the conditions of life over large areas of land and ocean and, on the other hand, as a result of the effect of antibiotics on humans, as well as the effect of strong chemical substances. In the near future we may also witness direct intervention in the genetic structures of life. Modern biology is not in a position to foresee all the consequences of these developments. For the time being, there is perhaps only one way to obtain an answer to these questions, and that is by carrying out direct experiments on biological systems that are subjected to unusual conditions. This could be done by extensively using the methods of the design of experiments, so as to appropriately plan research which simultaneously involves a large number of independent variables. But the trouble is that such experiments can be carried out only on a small scale (in space and time). Therefore, we cannot expect to obtain from them information about similar situations realized on a large scale, when low-probability events can manifest themselves that (as noted above) can radically change everything.

Perhaps we are now in a position to give a formal definition of life: living systems are systems which are random in their essential manifestation, but in this case the randomness (in contrast to inanimate nature) is such that extremely improbable events play a decisive role. Thus, we are very near not only to ontogeny and phylogeny but also to the creative manifestations of human activity. Modern science, including such branches as probability theory and mathematical statistics, is not yet ready to deal with these structures.

The Role of Computers in Efforts to Describe the World Scientifically[5]

The progress of computers has given rise to many an illusory hope. One of these is the conviction that it is possible to construct extremely complicated mathematical models based on an elegant analysis of crude experimental data. Computers were thought to be like "mathematical spectrographs" which possess an extremely high resolvency with which they decompose the experimentally observed data into the components which are not immediately observed in the experiment. It would be hard to indicate a paper where such an idea was explicitly formulated and seriously grounded. It seemed to emerge by itself and has been and still is shared by many. This is a *paradigm* brought forth by the unprecedented capacity for computational mathematics. But, sooner or later, any paradigm is subjected to revision.

---

[5] A brief version of this section was published in the journal *Industrial Laboratory* (No. 3, 1978).

I shall analyze the situation using the problems of chemical kinetics. The researcher observes time changes in the output of the final product in a chemical reaction and wishes to reconstruct the mechanism of intermediate reactions which are not immediately observed in the experiment. He proceeds from the assumption that on the basis of some prior knowledge he can write a system of differential equations which would reflect the mechanism of intermediate reactions and give the model parameters. The prior information is not formalized; it includes the researcher's previous experience and, probably, the whole progress of chemistry (which depends on the researcher's level). Prior knowledge is always subjective since it reflects the attitude either of the researcher himself or of the school to which he belongs.

A system of differential equations may be regarded as embodying initial axioms, on the basis of which one should decompose experimentally observed data into their components. The list of reactions is thermodynamically open. Their number seems to be wholly determined by the technology. At present, one may come across papers including up to twenty parameters. Some time ago, when computers were less efficient, the number of intermediate reactions used not to surpass two. I think it would be instructive to build sciencemetric graphs showing the growing number of intermediate reactions analyzed as computers become more and more intricate.

The tendency is, at least, obvious: the progress of computers gives rise to a temptation to create models whose complexity depends on the computing techniques. But since the progress of the latter may be practically infinite, the complexity of models should also increase infinitely.

We have by now accumulated a broad experience which points out the principal difficulties one faces when trying to use computational methods to decompose the initial experimentally observed data into components. The problem which arises while measuring radioactive decay is well known. In this case one deals with the function

$$f(x) = A_1 e^{-\lambda_1 x} + A_2 e^{-\lambda_2 x} + \ldots + A_m e^{-\lambda_m x}$$

where the parameters of decay λ_1 and activity A_1 are unknown. At first sight, the problem is similar to that of decomposing the oscillation process registered as a whole into separate periodic constituents, but the lack of orthogonality for the exponential functions results in difficulties unknown in harmonic analysis. These difficulties were long ago described in texts (see, e.g., Lanczos, 1956).[6]

---

[6] Lanczos (1956) gives an illustration of the case when observational results generated by the sum of three exponential functions proved to be well approximated by the sum of two exponential functions, and the decay parameters for one index decreased from 3 to 1.58, and for the other, from 5 to 4.55 while

One also has to resort to methods of "mathematical spectroscopy," analyzing nonadditive multicomponent systems according to their absorption spectra obtained by means of physical spectrographs (Vasil'ev, 1976). Here several essentially different problem formulations are possible. One of them is an analysis of multicomponent mixtures with partially known composition. In its ordinary form, the problem is solved by normal regression analysis approximating the absorption spectrum of the unknown admixtures by an algebraic polynomial; at every stage of selecting a model it is necessary to solve a system of linear equations with $n + S + 1$ unknown quantities, where n is the number of components analyzed and S is the polynomial's power. Its form turns out to be equivalent to a common least-squares method which, instead of covariance matrix, includes special matrices calculated beforehand, which possess the following projective properties:

$$\bar{\mathbf{c}} = (\tilde{\mathbf{k}}\mathbf{P}_i\mathbf{k})^{-1}\mathbf{k}\mathbf{P}_i\mathbf{D}$$

i.e., at every stage of selecting a model, one has to solve a system of linear equations with only n unknown quantities. Here, $\bar{\mathbf{c}}$ is a concentration column, $\mathbf{D}$ is an optical density column, $\mathbf{k}$ is a matrix of absorption coefficients, and $\mathbf{P}_i$ is a projection matrix built on algebraic polynomials up to the ith power inclusive. The effect of applying such a projection matrix is similar to that of a low-frequency filter which lets through only high-frequency constituents (with the number larger than i) in the expansion spectrum of a target function with respect to the given basis of algebraic polynomials in Hilbert space. All problems of this sort are fraught with troubles typical for a non-orthogonal regression analysis. For the component we are interested in, it is impossible to fix confidence limits which would not depend on the choice of all the other components. This results in the absence of a unique solution of the problem.

Of course, there are many cases when such troubles may be neglected. One of the principal problems of "mathematical spectroscopy" is to outline clearly its applicability in various real situations.

But let us return to constructing models in chemical kinetics. The process passes through the following stages. The system of differential equations is analytically integrated. There emerges the function

$$\eta = \varphi(\mathbf{x}; \Theta)$$

---

the amplitudes ratio was distorted from 1:2 to 1:7. In a simplified problem, when the number of components and the approximate value of every index are known and the amplitudes A_i and corrections to the approximate estimates λ_i are to be found, decomposition results are essentially better though they still remain disappointing. At worst, instead of the true value 1.0, λ gets the value 0.5, and it proves less adequate than the preliminary estimate with $\lambda_i = 1.2$.

nonlinear with respect to parameters. The task is to elaborate the optimal experimental design generating the matrix of independent variables $\mathbf{X}$, to realize it, and then, using the least squares method, to estimate parameters Θ, to find confidence limits for the estimates, etc. All this is naturally preceded by linearization. Assume that we confine ourselves to representing the function by a Taylor polynomial in the neighborhood of the point Θ_0. In this case, while designing an experiment and evaluating its results[7] we shall have to deal with the covariance matrix $(\mathbf{X}^x\mathbf{X})^{-1}$ obtained from the independent variables matrix $\mathbf{X}$ of $N + k$ dimension (N is the number of experiments, and k is the number of parameters estimated)

$$\mathbf{X} = \{x_{r,u}\}$$

where the element of the matrix

$$x_{r,u} = \left[\frac{\partial\,\varphi(\mathbf{x}_u;\,\Theta)}{\partial\,\theta_r}\right]_{\Theta\,=\,\Theta_0}$$

is a partial derivative with respect to the parameter θ_r in the point $\Theta = \Theta_0$ under the fixed values $x_{1u}, x_{2u}, \ldots, x_{ku}$ corresponding to conditions of the uth experiment.

The optimality criterion for an experimental design may be formulated as follows. Points in the space of independent variables should be situated so that the determinant $|(\mathbf{X}^x\mathbf{X})^{-1}|$ will be minimal. The functional determinant (a Jacobian) $|(\mathbf{X}^x\mathbf{X})^{-1}|$ gives the transformation of coordinates of experimental space (the space with coordinates $\eta_u = \varphi(\mathbf{x}_u;\,\theta)$ into the space of parameters. The minimal Jacobian would mean the approximately minimal volume of a k-dimensional ellipsoid in the space of parameters determining confidence limits of the parameters. Design efficiency depends here on the values of partial derivatives. Therefore, designing an experiment for functions nonlinear with respect to parameters requires knowledge of preliminary parameter estimates. One can apply sequential methods of experimental design when all the research is divided into separate stages, parameters are estimated for every stage, and these estimates are used as a new preliminary approximation for the next stage. The optimal design is each time calculated anew. But the calculations may prove so complicated that they will reduce to nothing the gain from the optimal sequentially improved design.[8]

---

[7] I remind the reader that under matrix denotation the vector column Θ is given by the relation $\Theta = (\mathbf{X}^x\mathbf{X})^{-1}\mathbf{x}\mathbf{y}$ where $\mathbf{y}$ is a vector column of observational results.

[8] Designs minimizing the volume of the variance ellipsoid of parameter estimates are called D-optimal. Despite its obvious nature, the criterion does not completely determine behavior of parameter estimates: the volume of the ellipsoid may be minimal, but the latter may be too stretched along an axis. If the

If there exist several alternative models capable of describing the mechanism of the phenomenon, discriminating experiments may be carried out to select the best one. A new problem formulation generates new optimality criteria. One of them is based on a measure depending on the difference between sums of square deviations. For two rival hypotheses, after carrying out N observations we shall deal with the difference of two values

$$S_j(N) = \sum_{i=1}^{n} [y_i - \eta_j(\mathbf{x}_i, \Theta_j)]^2 \qquad (j = 1, 2)$$

If the difference is not large enough to give preference to one of the rival hypotheses, it is suggested that the next $(N + 1)$th experiment should be held at the point where it is expected to reach its maximal value.

The second criterion is based on the use of the modulus of the logarithm of the generalized likelihood ratio calculated for two rival hypotheses: measurements are situated so as to achieve the most rapid increase of the value. Last but not least, the third criterion exploits, as a measure for discriminating hypotheses, the Kulback measure of divergence, well known in information theory: the choice of an optimal design consists in maximizing the value. A dual problem formulation is also possible: a design may be chosen with the goal of either estimating parameters or discriminating methods. The optimality criterion is here based on the necessity to minimize a weighted sum of two addends, each

---

researcher wishes to minimize the maximal axis of the ellipsoid of variance, he should build a design matrix which would have a corresponding covariance matrix with a minimal eigenvalue. This will be the so-called E-optimal design. The researcher may require the average variance of parameter estimates to be minimal. To this requirement corresponds an ellipsoid with the least sum of square axes length. Corresponding designs are called A-optimal; the respective covariance matrices have a minimal trace value. The list of criteria is easily continued. If we deal with simple polynomial models where sequential designing is irrelevant, we can build designs to meet various criteria, estimate them from the standpoint of other criteria, and then find a solution satisfying various approaches. For details see, for example, Korostelyov and Malutov (1975). When models are nonlinear with respect to parameters, it is wise to follow the course of sequential designing, but in this case computation of admissible designs (with respect to criteria) becomes an extremely complicated task. So one has to give preference to a certain criterion. The highly readable review by Kafarov et al. (1977) briefly discusses the question of selecting a criterion in problems nonlinear with respect to parameters; e.g., attention is paid to the fact that the D-criterion, descreasing significantly the volume of the confidence ellipsoid, does not change correlation coefficients and, therefore, does not improve the ravine surface (unsmooth surface without a distinct extremum) of the sum of square deviations. It is also of interest that a number of illustrations have shown that the volume of a hyperellipsoid of the confidence region decreases mainly at the expense of compressing it along the minor semi-axes, which are smaller than the major ones in the same order of values. At the same time, the aim of the E-optimality criterion is to make the confidence region close to a sphere. But all these observations made by Hosten for separate cases are, naturally, insufficient to allow an ultimate conclusion for a general case. In selecting an optimality criterion, one should learn to take into account the type of a model.

of them being responsible for one problem formulation. The difficulties of problem formulation are then shifted to the choice of weights. [The reader is referred to Nalimov (1971) for a brief and popular description of the questions concerning the choice of criteria in discriminating a problem and to Fedorov (1972) for a detailed and rigorous account.]

Now we pass to considering confidence limits for parameters in the models with nonlinear parametrization. The problem of confidence limits becomes crucial since parameters are regarded as physical constants inherent to the mechanism of the phenomenon described by the model nonlinear with respect to parameters. The constants have an unambiguous physical interpretation (counterparts) and their numerical estimates are to be stored in reference books or in the memory of computers operating in informational systems.

It is exactly at this point that practically insurmountable difficulties arise which often turn the whole process of parameter estimation into an illusory activity. Having analyzed the process of studying phenomenological mechanisms by mathematically expanding the total data according to the parameters of a hypothetical model, one can discover three sources of uncertainty. The first one is the random error in measurements, the second stems from the fact that model parameters cannot be estimated in a unique way, and the third source lies in the possibility of giving diverse formulations of the chemical axiomatics determining a set of several intermediate reactions.

I would like to start the discussion with the first type of uncertainty. The essential thing is that the estimates of some parameters correlate highly (their correlation coefficients reach sometimes .99 or even .999). In such models, applying an experimental design to orthogonalize the information matrix proves futile. In this case it is of no use to record confidence limits as the familiar relations $\hat{\theta}_i \pm 2\sigma\{\hat{\theta}_i\}$, since such a step ignores the information concerning correlation of estimates. When looking through a reference book containing results of other studies, the researcher, using the theoretical data at his disposal, may wish to accept as a parameter estimate any value situated close to the boundaries of the confidence interval. In accordance with statistical theory the researcher has every right to act this way and to choose at will within the given interval of confidence. But because of highly correlated estimates, he would immediately have to recalculate confidence intervals for all other parameters. To make this possible, reference books would have to include, together with parameter estimates, the corresponding covariance matrix, which is very cumbersome. Moreover, the reference book as an instrument of visual data presentation loses any utility because one has to resort to a computer. What can be done to preserve a visual type of presentation? Even if eigenvalues of the covariance matrices which gave the axes of the variance ellipsoid of estimates are included in the reference book, presentation of a multidimensional ellipsoid will not become more visual. It is also possible to circumscribe a multidimensional parallelepiped around the ellipsoid. The presentation will acquire a more visual character since every parameter will be given maximal confidence limits independent of how the ultimate values for other parameters are fixed. But that will cost us dear since in a multidimensional case the volume of the parallelepiped will be several times larger than that of ellipsoid, especially because in a nonlinear parametrization we deal not with ellipsoids but with an ellipsoid-like shape which in a two-dimensional section looks oddly banana-like. At last, we may introduce in the reference book two-dimensional sections of ellipsoid-like figures, though such data will be too bulky. For a problem with 20 parameters, the number of sections equals 190, and each of them should be accompanied by a development.

So it is obvious that, in the case of a nonlinear parametrization, the uncertainty stemming from the experimental error is not at all easy to present visually. And this is only one source of uncertainty.

Now let us consider the second source of uncertainty: lack of a unique computation resulting from the complexity of calculation procedures. Here we should keep in mind several factors. (1) Linearization procedures may be varied: a Taylor polynomial may be given differently — once on the basis of only the first derivative, another time introducing the second derivatives as well, etc. Linearization may also be achieved without expanding into Taylor series but for each model choosing an expansion according to some functions natural for it. (2) When estimating parameters, we can get different results because an information matrix proves poorly conditioned, i.e., its determinant is close to zero, as a result of the high correlation of several parameters. (3) Computations may yield different results because the initial parameter estimates have been chosen arbitrarily. All iteration procedures require selection of the initial point. If, as usually happens, besides the absolute minimum, there exist a lot of others, poorly chosen initial estimates may lead to convergence in an undesirable stationary point on the surface of the sum of squares, or there may be no convergence at all because calculation results will spread on the surface where the stationary point is being searched for.

The well-known book by Draper and Smith (1966) describes the troubles connected with a linear least-squares method (after linearization of the model by expansion into a Taylor series) considered in the sequence of its stages in the following manner. The computation procedure may converge very slowly; strong oscillation with partial increase and decrease of the sum of squares may arise, though, in the long run, the solution may become balanced. As a matter of fact, the procedure may not converge or it may even diverge so that the sum of square deviations will grow with each iteration, though it is possible to show this method to be always convergent (the rate of convergence depends on the model whose parameters are estimated; the process may also converge slowly, and with strong oscillation). For more details, see also Bard (1974). If the computation procedure selected to solve the problem does not converge, generally speaking, the researcher knows how to act. He may change the linearization procedure[9] or, having rejected it altogether, he may change the computation procedure; within the frame-work of any such procedure, initial approximations may be varied, and this process can be formalized to a certain degree by selecting a combinatorial lattice on the basis of experimental design. Thus, for example, initial approximation for each parameter may be varied at two levels, and then points in the factor design (k is the number of parameters) or its regular replication will be the nodes of the lattice.

In a sequential design, when the initial approximations at the $(i + 1)$th step are the values obtained at the ith step, the role of the initial approximation received from the first rough preliminary experiment seems to decrease, but it hardly is eliminated for the squares sum surface of any type.

Sometimes the researcher has to take drastic measures — either to reparametrize the model or to record it in an absolutely nontraditional measure, the way it is done by Gontar' (1976). Sometimes one can manage to unite two highly correlated parameters, modifying the model only slightly so as not to violate its physical meaning. In any case, it has been noticed that a slow convergence takes place when the contour curve for the sum of square deviations has a shape of stretched bananas.

All this is well described in numerous books, including the book by Draper and Smith (1966) mentioned above. And the research is always assumed to make the best and the only right decision. But whence does this come? Not only the answer to this question is lacking — the question itself is not asked. We dare claim nowadays that, in the problems of non-

---

[9] Expanding the function with respect to the powers of independent variables may also be considered as linearization. This is equivalent to presenting it as a Taylor polynomial with coefficients

$$\beta_1 = \frac{\partial \varphi}{\partial x_1}, \; \beta_2 = \frac{\partial \varphi}{\partial x_2}, \; \ldots \; \beta_{12} = \frac{\partial^2 \varphi}{\partial x_1, \partial x_2}, \; \ldots \; \beta_{11} = \frac{\partial^2 \varphi}{\partial x_1^2}, \; \beta_{22} = \frac{\partial^2 \varphi}{\partial x_2^2}$$

which get numerical values after the experiment. In contrast to the case considered above, here, during linearization, derivatives are taken with respect to independent variables and not to parameters. If the researcher is satisfied with approximating the function by a first-order polynomial, then all regression coefficents can be estimated with zero correlation coefficients. All troubles connected with parameter estimation and their confidence limits disappear. But the initial parameters of the model disappear too; they are replaced by pseudoparameters, that is, regression coefficients. Hence it follows that *the less we wish to learn, the more definite becomes our knowledge* (the experimental potentiality being the same).

linear parametrization, statistics has lost one of its most attractive merits. It stopped being at the same time a science setting the rules for parameter estimation and a metascience estimating the reliability of its results. In any case, confidence ellipsoids, even if they could be visually presented, reflect the uncertainty stemming from random error and in no way reflect the uncertainty related to nonstandard computational procedures. One and the same problem, nonlinear with respect to parameters, can obviously yield essentially different solutions in different and equally good computational centers. Moreover, provided a good level of critical attitude, it is possible to obtain different results within one center.

The uncertainty stemming from nonstandard computational procedures, as a matter of fact, may be estimated in a purely statistical manner using, for example, variance analysis with hierarchical classification. The lowest level would contain the uncertainty due to choosing initial approximations, the highest level would be occupied by the uncertainty related to choosing models during their reparametrization, and the intermediate levels, by all other nonstandard computational procedures. But the whole structure would have a cumbersome appearance. It seems rational to limit ourselves merely to presenting observational results in several versions — naturally, only in those which a computer–mathematician thinks fit for this purpose.

If we look through journal publications concerning models nonlinear with respect to parameters, we shall notice an amazing disregard of uncertainty estimates. A covariance matrix is only very seldom given as a whole. Usually its diagonal elements are given, and this, as I have already mentioned, would be correct only for an absolutely orthogonal information matrix. As a rule, in such papers one will not find anything about the uncertainty due to computational procedures. A welcome exception is the article by Korostelyov and Malutov (1975), but despite their criticism the authors present observational results by a single model.

And now we shall at last analyze the third source of uncertainty, initial chemical axiomatics. I remarked above that the list of differential equations giving the mechanism of intermediate reactions is thermodynamically open, and the researcher selects his own rationally limited list of possible intermediate reactions based on the micro-paradigm of the scientific school to which he belongs. A researcher with a critical mind may suggest several such lists, and this accounts for the problem of discriminating hypotheses mentioned above. A complicated apparatus for discriminating procedures implicitly introduces a new source of uncertainty into results of investigations.

The first thing to be noted here is that we are not (and generally cannot be) sure that the rival models include the "true one." But without this assumption, the formulation of a discriminating problem proves groundless. Discriminating procedures will yield different results if the sets of hypotheses to be discriminated vary. In any case, the problem of a discriminating procedure converging to the "true model" no longer has any sense.

The second point is that a discrimination procedure may be wholly determined by the arbitrariness in parameters estimation introduced at the stage of computation. Recent experience has shown that not all models selected in the laboratory prove fit to describe the corresponding processes at a plant, though formally the models for all processes have been selected with equal rigor.

The third feature is that a model as a whole is variant with slight modifications in the initial chemical axiomatics. Such a modification of the list of intermediate reactions results not only in the appearance of new model parameters but also, as a consequence of high internal correlation among parameter estimates, in the change of numerical values of the parameters characterizing the unmodified reactions. It is practically impossible to try all possible combinations of intermediate reactions, and besides, again due to high internal correlation, extensive change of numerical values may lead to models which prove almost the same when compared with experimental results in a fixed interval of independent variable values.

The fourth peculiarity is that extrapolations are not correct. The very essence of chemical problems often enables discrimination of hypotheses in a narrow and easily achieved interval of variation so that further significant information could be obtained by extrapolating the best model of those selected. Such an approach is hardly correct since it follows from what has just been said that, while discriminating, we only estimate the interpolational power of a model, and a model whose interpolation properties for a narrow interval are quite good may at the same time be of rather poor extrapolation power. Some examples borrowed from practice show that the best model for extrapolation has been that estimated as the worst one in a discriminating experiment carried out as rigorously as possible.

The fifth point is a question. If, discriminating in a narrow interval, we find a model to behave not in the best possible way, is this enough to reject it? A slight modification of the initial chemical axiomatics might weaken the interpolation power of the model, but this is not to say that it does not reveal certain changes in parameter estimates (again due to their high correlation) which may come out in a wider but practically unrealizable range of independent variables.

Here we may end our list of claims to discriminating procedures. Now the question can be formulated in a more general philosophical form: Whence comes the whole theory of discrimination? The answer is very simple: in the terms of Kuhn (1970*a*), its source lies in the existing paradigm, which states that the world, on the one hand, is arranged so that everything within it is governed by the only possible laws of nature and, on the other hand, it also possesses such a property that a scientific experiment allows us to discover these laws.

The first part of the paradigm has begun to lose its supporters in modern physics. It is opposed by the "bootstrap" philosophy (Chew, 1968; Capra, 1976), which holds nature to be an interrelated dynamic web irreducible either to elementary blocks of substance or to fundamental laws, equations, or principles. The term "bootstrap" cannot be related to a single model; it can only be applied to a combination of internally consistent models among which none is more fundamental than the rest. Consider the multitude of models in the physics of elementary particles (Moravcsik, 1977). The past 20 years have witnessed the emergence of numerous theories and models, some of which are conceptually contradictory of others. None of them can be rejected because each explains a part of the observed phenomena and none of them can be accepted as the only one because none can explain everything. I draw the reader's attention also to a very interesting paper by Smirnov (1977*a*) in which the problem of plurality in models in physics is discussed.

As to the second part of the paradigm, after the well-known work of Karl Popper (1963, 1965; see also Chapter 1), it has also become philosophically clear that the role of an experiment in science is limited: a hypothesis can never be experimentally supported. The only thing that can be done is to show that the experiment does not contradict the hypothesis. But the same experiment may prove consistent with some other hypothesis, as yet unformulated, and a new experiment carried out to confirm a new theory may become crucial for a hypothesis consistent with previous experiments. Any hypothesis not refuted by an experiment remains open to further tests, and, according to Popper, here lies the source of progress of natural sciences.

But if a hypothesis is refuted by an experiment, is it always rejected immediately and unconditionally? Above, in Chapter 1, I borrowed an example from Monod (1975) which described an awkward situation with Darwin's theory. Thomson (Lord Kelvin), the physicist, demonstrated by means of exact calculations that solar energy could not suffice for the evolution of life on the Earth. Darwin was depressed by these calculations. A direct experiment — measurements of heat received by the Earth, of the dimensions of the Sun, and of fuel caloricity — came to contradict his theory. However, the latter was not rejected. Monod remarks that at present we may state that Darwin's evolutionary theory implicitly contained the concept of solar nuclear energy though nobody could have had such an idea at that time. Besides, adds Monod, Darwin's theory also implied the concept of a discrete biological code, contrary to Lamarck, who assumed the continuity of heritability.

One can also cite numerous examples of an opposite kind which demonstrate that negative results have been of the utmost importance for the development of science. One of them is the famous experiment of Michelson and Morley, which gave an impetus for a new era in physics.

From all these historical contradictions, it obviously follows that negative experimental results acquire significance only when combined with a system of meaningful reasoning. Statistical methods of model discrimination as they are presented in books on experimental design are too formalized. They exclude a meaningful discussion of experimental results, and at the same time they prove inconsistent as a result of their formalism since they ignore the uncertainty caused by the computational difficulties generated by the structural peculiarities of the models. But this has already been discussed in detail.

Thus, the system of our initial concepts should evidently be modified. The researcher's new paradigm must not presuppose the existence of the one and only true model, even if he investigates the mechanism of phenomena. Why should the researcher assume the existence of what proves illusory in the long run? He might better follow the example of the physicists who, in quantum mechanics, abandoned Laplacian determinism and even its weaker forms after the illusory nature of its serviceability became obvious.

Besides, the results of studying the mechanisms of phenomena should be presented not by a single model but by several of them. The variety of models may result both from the different initial chemical axiomatics and the insurmountable absence of unique computational procedures. It seems rational to acknowledge the possibility of performing discriminating experiments since they are rather informative, but they should be given only a limited significance.

If we decide to hold this viewpoint, we shall immediately contradict the traditional view on the role of mathematical statistics in research. Like Ronald Fisher, I long believed that the task of statistics was the reduction of data. A statistically trained researcher is able to present his results in a much more compact form than if he registered experimental data directly as they were obtained. Under a new problem formulation, *mathematical statistics* will be used not to *reduce* data but to *unfold* them. Numerous models nonlinear with respect to parameters with their confidence limits development of two- or three-dimensional sections of ellipsoid-like figures will look more complex and cumbersome than the immediately observed values — an independent variables matrix X and a vector of observational results Y.

But will the researcher then be able to perceive the information about the process he studies when it is presented in such a cumbersome way? Imagine an audience looking at the screen and seeing there the filmed variety of graphic data accompanied by certain comments. The researchers will watch all these data extracted from the experiment while varying both computation procedures and initial chemical axiomatics. This process may also be presented as a human–computer dialogue. The important thing is whether the researcher has an insight that will allow him to present the mechanism of the phenomenon in a new way and to outline further research. In other words, whereas the reduction of data used to be performed on the logical level, the application of statistical procedures has transferred it to the intuitive level. A computer unfolds the information contained in the experiment; a researcher will have to reduce it while comprehending it theoretically. The potentialities of a computer meet some unexplored human potentialities, and they switch roles: humans are now to meditate over the free information flows generated by computers.

It seems pertinent to draw an analogy with what is now happening with the foundation of biology, the theory of evolution. At present there exists a set of evolutionary theories which are hard to classify. They may be said to contain as a basis a list of "evolutionary factors." The variety of theories is formed by ascribing weights to these factors. Theories with similar factor weights are naturally unified into groups, and these groups are given a label. The mechanism of constructing evolutionary theories may be called "logical spectroscopy." A biologist or, to be more correct, a paleontologist observes the sections of the distant past and attempts to decompose them according to all evolutionary factors by ascribing weights to the latter. The beginning of our century, up to the 1920's witnessed the emergence of novel evolutionary theories. The process has by now slackened or stopped altogether. A theoretical biologist behaves in the following manner: after becoming familiar with the whole variety of evolutionary theories, he constructs his own, i.e., ascribes new weights to the set of known evolutionary factors. As a result, we have as many theories as there are theoretical biologists. This variety, individually reduced to compact homogeneity, opens up the possibility for individual, creative work. [The ideas concerning the theory of evolution are stimulated by an extremely interesting report, "Classification of Evolution Theories," made by S. V. Meyen at the School of Young Scientists in Theoretical Biology in Kondopoga, February 1977. See also his paper (Meyen, 1975).]

I would like to conclude this chapter with the following considerations.

One of the principal scientific tasks is to explain observational results according to the underlying factors (or mechanisms). The vast data accumulated up to the present show that this task in its general formulation cannot be unambiguously fulfilled either by the methods of "numerical spectroscopy" or by those of "logical spectroscopy." Computers have only made the task more difficult. But now a new way seems to emerge: *reducing information by means of unfolding it,* i.e., presenting it through a set of models. However, this may be another illusion.

A few words must be said about another approach to simulating complex systems by computers, which does not refer to "mathematical spectroscopy." I have in mind the grandiose program of simulating five ecosystems: the desert, coniferous and foliage forests, the tundra, and the prairie, carried out in the United States in 1969–1974. Expenses for the research of only the three latter systems exceeded 22 million dollars, 8.6 million of which were allotted directly for the simulation, synthesis, and control of the whole project; 700 researchers and postgraduates from 600 U.S. scientific institutions participated in the project; 500 papers were published by 1974, though the final report is not yet ready. Mathematical language was used in the project to give an immediate (not reduced) description of the observed phenomena. A lot of different models were used which were divided into blocks with an extremely great number of parameters (their total number reaching 1,000), and yet it was emphasized that the models described the system under study in an approximate and simplified manner. The researchers had to give up any experimental verification (or falsification) of models; instead they used "validification," which means that a model is accepted if it satisfies the customer and is particularly favorably evaluated if bought by a firm. At present, all these activities are being evaluated. Mitchell et al. (1976) conducted a thorough analysis of the material and evaluated the simulation of the three ecosystems in an extremely unfavorable way. From a general methodological standpoint, the following feature is important to emphasize: the language of mathematics, for the first time, is allowed to unify different biological trends, and this has happened without a generally novel or profound understanding of ecology. Mathematics was used not to reduce complexity but to give a detailed, immediate description. This is a new tendency in science. But where will it lead? The time is not yet ripe for a final conclusion, but scepticism is quite in order.

Concluding Remarks: Dialectics of Reduction and Expansion of Knowledge in the Development of Science

We are unintentionally witnessing an amazing phenomenon. Up to now, the development of science was directed at obtaining a form of knowledge with maximum reduction of redundancies—a compact representation of the world. But now virtually before our eyes, there appears a tendency to present knowledge in an expanded form, by a multitude of equally legitimate models. Each individual model implies the presentation of knowledge in reduced form, but the acknowledgment of the legitimacy of many models describing one and the same phenomenon allows the presentation of our knowledge of the world in an expanded form. This is where the dialectics of the development of science is manifested.

The necessity to resort to many models first became evident long ago in the problems of multidimensional statistical analysis. The passage to the principal components is definite if the metric of the initial variables is given, but it may be given in various ways. Factor analysis is indifferent toward the initial metric, but the rotation of axes, obtained after a corresponding transformation, is performed arbitrarily (for more detail, see Nalimov, 1971). In non-orthogonal multidimensional regression analysis, the same experimental results may be equally represented by the entire diversity of models (an instance of such analysis is given in Chapter I of Nalimov and Chernova, 1965).

From the general methodological point of view, though, it was possible to ignore these facts, by assuming that these are merely particular cases related to situations without sufficient initial theoretical premises. However, the contents of the above section make the matter look more complicated. Even if we are provided clearly formulated initial theoretical premises of the mechanism of phenomena, we still have to resort to many models. Moreover, if we compare all this with the prevailing situation in the physics of elementary particles, the problem of an expanded representation of knowledge will acquire a threatening dimension.

Chapter 9

The Penetration of the Humanities into Other Fields of Knowledge[1]

Reflection on the Ways in Which Science Develops

> *"Cheshire-Puss," she began, rather timidly, "would you tell me please, which way I ought to go from here?"*
> *"That depends a good deal on where you want to get to," said the Cat.*
> *"I don't much care where," said Alice.*
> *"Then it doesn't matter which way you go," said the Cat.*
> *"So long as I get somewhere," Alice added as an explanation.*
> *"Oh, you're sure to do that," said the cat, "if you only walk long enough."*
>
> LEWIS CARROLL

Penetration of the Humanities into Scientific Disciplines

If we observe closely the process of the development of science, side by side with the penetration of mathematics into the humanities we see the humanities penetrating into the fields of knowledge which traditionally they did not enter.

However, this latter process goes on in quite a different manner. When mathematics penetrates into non-mathematical studies, it becomes the

[1] This chapter was published in Russian in the journal *Znanie-sila* (No. 5, 1979) under the title "Yearning for the Lost Integrity."

199

language in which the models are built, problems are formulated, and decisions are taken, but the problems and conceptions basically do not change. In contrast, when the humanities penetrate into other fields of knowledge, they turn into branches of these fields, thereby enriching and deepening them. In penetrating into foreign branches of knowledge, the humanities start to lose their speculative, descriptive character. Their all-embracing and therefore inevitably vague ideas begin to turn into precise logical constructions. Their role in the analysis of observable facts and of quantitatively estimated parameters increases sharply.

The process of "humanization" of knowledge, as well as that of its mathematization, started a very long time ago, but it became quite prominent only recently. The process advances sometimes rather painfully. The humanists often resist desperately, unwilling to allow branches of knowledge that have belonged to them from time immemorial to acquire a quite unfamiliar precision and, therefore, narrowness in both the formulation and the solution of problems. Scientists, too, are far from being always ready to accept concepts which might broaden their thinking if they are introduced by the humanities. It is not so easy for a natural scientist to agree that one must not only study a concrete branch of knowledge but must also think of its logical and sometimes its purely psychological foundations.

The most remarkable aspect of the humanization of knowledge seems to be the acknowledgment of the significance of man and the peculiarities of human thinking for the evolution of our knowledge.

Let us now trace in detail the process of the penetration of the humanities into some other fields.

Statistics.   The first use of the word "statist" was in fiction (Yule and Kendall, 1950). It appeared in *Hamlet* (1602), in *Cymbeline* (1610–1611), and in *Paradise Regained* (1710) and its meaning in these works is rather vague. The word probably stems from the Latin word "Status," which means "a political state, or condition." Later, the term "statistics" appeared in science. One can trace the evolution of the term with the help of a collection of its definitions (to be more accurate, of statements made about this term) from the middle of the eighteenth century up to the present (see Nalimov, 1981). Roughly speaking, three major stages in the evolution of the term's meaning can be outlined.

At first, the term meant the teaching of the economic and political conditions of a state, based on the analysis of quantitatively expressible economic factors. Among the earliest definitions of the term statistics were the following: "the science that teaches us what is the political arrangement of all the modern states of the known world" (Bielfeld, *Elements of Universal Erudition*, 1770); "a word lately introduced to express a view or survey of any kingdom, country, or parish" (*Encyclo-*

"Humanization" of Knowledge

paedia Britannica, 3rd ed., 1797); "the most complete and best grounded knowledge of the condition and development of a given State and of the life within it." (C. A. v. Malchus, *Statistik und Staatskunde*, 1826). Within this frame of reference, statistics was a purely humanistic subject (see also Shelestov, 1977).

At the second stage, statistics came to signify the processing of any quantitatively presented data without regard for their source, be it in socio-political and economic research or in the natural and technological sciences. By and by statistics was changing from a science describing the conditions of a state into a methodological science for processing data. Mathematical statistics appeared, and a purely humanistic discipline became mathematical. At this stage some authors made an attempt to draw a demarcation line between mathematical statistics and statistics as a social science.

At the third stage of its evolution, statistics is sometimes defined very broadly — as a metascience. Its object is the logic and methodology of the other sciences — the logic of decision making and the logic of experiment in them.

Probably the most important feature is that many scientists working as statisticians have abandoned the term "mathematical statistics." Indeed, any manipulation of data ought to rely upon mathematical treatment. A mathematically structured discipline serves to solve non-mathematical problems. And if the philosophy of science is considered a part of the humanities, the evolution of statistics illustrates the humanization of a mathematical discipline.

Mathematical statistics should answer the question of what a good science is. This question is posed by man, and the answer to it should satisfy him: it should correspond to what is considered scientific within science or, to be more precise, to the system of scientific prejudices — because it is in this way that the scholars and scientists of the future will evaluate our conceptions of what is scientific in the science of today.

Logic.   Imagine an intellectual of the beginning of the twentieth century. What did the word "logic" mean to him? He would, indubitably, instantly recall the Aristotelian papers collected under the title of *Organon*. However, this was nothing more than a codification and systematization of rules of reasoning with which reasonable minds are quite familiar. Then he would probably remember Thomas Aquinas, who endowed logic with an ontological character, and Francis Bacon's *Novum Organum*, in which deductive logic was opposed by inductive logic, and Hume, who was the first to demonstrate the impossibility of deductively grounding inductive logic (or, in modern terms, the logic did not yield to expression by an algorithm which conceals creative thinking). Everything connected with logic gave an impression of something obsolete and useless — "school logic," a branch of knowledge fruitless through a twenty-century period. In Russia only the curricula of "gymnasia"[2] included a course in logic; no such course was studied in "realistic schools" whose graduates entered higher technological institutions. Indeed, this was a subject with no prospects for a pragmatically oriented intellectual.

---

[2] Before the revolution, the system of education in Russia was rather complicated. It had several types of high schools, two of which were the so-called "gymnasia," with a classical direction in education, and "realistic schools," orienting their pupils toward the exact sciences.

In the 1930's formal logic was considered in the Soviet Union as an ideology hostile to Marxism. The article "Formal Logic" in the 1936 edition of the *Large Soviet Encyclopaedia* reads:

> Turning abstract definitions of thinking into the absolute and separating form and content, formal logic is always prone to idealism. . . . However, though formal logic does not reflect reality correctly, actually this is a teaching not only of forms of thinking, but also of forms of being — true, of being not real but distorted by a metaphysical mode of thinking. . . . Formal logic is the lowest step in the progress of human cognition and the stage of revolutionary class struggle of the proletariat for the overthrow of obsolescent capitalist society. In the hands of a counter-revolutionary bourgeoisie, formal logic became a reactionary tool for advocating the absolute. . . . Formal logic is a theoretical weapon of our class enemies. . . . We do not and cannot have any compromises with formal logic. . . . Formal logic is a one-sided exaggeration of some relative features of cognition into the absolute. . . . Formal logic is a methodological basis for anti-Leninist deviations in the Communist Party of the USSR.

But what is one to make of an advertisement of the firm *Logica Limited* published in the *Times* of London in 1971? Is it really a commercial firm whose activities correspond to its name?

I hope it will not be too much of an oversimplification to state that from the time of Aristotle up to the middle of the nineteenth century logic, remaining a purely humanistic subject, was in a torpid state. Boole's papers gave it a new life. Logic started turning into a mathematical subject; it became the basis for mathematics on the one hand and a purely applied discipline on the other. At the same time, intensive and fruitful development of properly logical problems, i.e., problems which retain a clearly humanistic flavor, was going on.

Mathematical logic is clearly a mathematical discipline, but its problems are rooted in traditional logic, which is without question a humanistic subject. We observe here not the mathematization of the humanities, but the humanization of mathematics, since a new mathematical discipline is being created whose task is to solve problems which had once belonged to the humanities.

It is interesting to see what place logic now occupies in our system of education. Many universities in the Soviet Union have two chairs: a chair of logic in the department of philosophy and a chair of mathematical logic in the mathematics (or physics–mathematics) department. As a matter of fact, their object of research is the same, but the departments approach it quite differently.

Psychology. Psychology seems to have lost its independence only recently. Part of its problems seem to have been taken over by philosophy; another part was placed in departments of physiology dealing with the higher nervous system. In the USSR[3] there was a time when psychology simply ceased to be an independent subject: there were neither higher educational institutions which turned out psychologists nor scientific centers carrying on broad psychological research. The Institute of Psychology of the Academy of Sciences of the USSR was organized within the Philosophy and Law Division only in 1971. Its most interesting feature is that the scientific–methodological control of the Institute is fulfilled by two divisions: that of Philosophy and Law and that of Mechanics and Control Processes. This complicated structure with a double subordination brilliantly reflects the process of regeneration of psychology, its transformation from a purely humanistic subject into a cybernetic one.

If we now consider the program of research of this Institute (Shishkin, 1972), we find there, side by side with traditional trends, such problems as human activity in the tracing regime, psychological mechanisms of decision making of a man–operator, memory mechanisms, space orientation of man, human interaction in the process of group activities, and human psychic states in extremal conditions.

It is the scholars with engineering interests who feel the acute necessity of solving the above-mentioned problems – those who are aware of the need to create not only computers but also something greater: the "man–computer" system. This became especially obvious after the development of electronic computers and space machines. The new aspects of psychology therefore acquired a new title (not too apt): engineering psychology. Hence started the regeneration of psychology, but not as a humanistic discipline in the traditional meaning of the word. Here again, an engineering discipline has to solve humanistic problems.

Linguistics.   Linguistics, one of the most ancient sciences, is also losing its humanistic appearance. Mathematical linguistics has been created, which, in the manner of Bar-Hillel, can be divided into statistical linguistics dealing with the frequency analysis of symbol systems and structural linguistics dealing with constructing abstract models of language. If the first branch can be regarded as a result of the mathematization of science, the second results from the humanization of mathematics: problems emerging within the humanities are formulated in the frame of mathematics. At any rate, Chomsky's theory of context-free languages is clearly a mathematical subject generated by linguistic problems.

---

[3] At the source of Soviet psychology stood the outstanding personality L. S. Vygotsky (1896–1934), the author of "cultural–historical" theory. His works contain studies of child psychology, psychotherapy, psychology of art, interconnection of language and thought, etc. Two of his books which have been translated into English are *Thought and Language* (MIT Press, Cambridge, Mass., 1962) and *Mind in Society, the Development of Higher Psychological Processes* (Harvard University Press, Cambridge, Mass., 1978).

Linguistics has also acquired some purely engineering aspects. The problems of machine translation, working out languages for computers, and especially the problem of "man–computer dialogue" have added engineering features even to such a purely humanistic field as semantics, though the principal problems of semantics have retained their humanistic core.

Economics.   Economics, too, is mathematized and enriched by using mathematical methods, e.g., those of mathematical statistics. But side by side with this process, another one is going on: quite a novel, purely mathematical branch of knowledge is being created, and only problem formulations are derived from traditional economics. Who knows what will happen next? Will economics preserve the humanistic nature of its system of judgments and use mathematics only as a sauce or will it be transformed into a profoundly mathematized, i.e., strictly deductive, subject?

At present (at least in the United States) econometrics has turned into a broadly developed deductive science only slightly connected with the analysis of actually observed phenomena. Here is what V. Leontief, an outstanding American economist, writes on this point:

> . . . economists, particularly those engaged in teaching and in academic research, . . . can demonstrate their prowess (and, incidentally, advance their careers) by building more and more complicated mathematical models and devising more and more sophisticated methods of statistical inference without ever engaging in empirical research. Complaints about the lack of indispensable primary data are heard from time to time, but they don't sound very urgent. (Leontief, 1971)

This is certainly something more than mathematization of traditional economics: this is the creation of mathematical economics as a deductive science, an independent branch of mathematics. Although these activities are perceived by many as performed for prestige, probably they will generate truly meaningful structures. This has happened more than once.

Management Science.   This word combination still sounds unfamiliar to the Russian ear. We have no departments turning out such specialists. But if one scans advertisements in English and American journals, one frequently comes across requests for these specialists. This is a mixed profession. It requires traditional engineering knowledge and also training in information theory, engineering, psychology, sociology, econometrics, mathematical statistics, and theory of decision making under conditions of insufficient knowledge. Here again, traditional engineering activities such as control of industrial enterprises are filled with humanistic content.

Philosophy of Science. The twentieth century has witnessed the emergence of a new branch of philosophy, which is becoming part of the natural and exact sciences. This discipline seems to have stemmed from the work of Russell, who studied the paradoxes of the theory of sets. Later, Hilbert (a mathematician, not a philosopher) set out to investigate the problem of proving the absolute consistency of mathematical structures, but he and his supporters failed, and in 1931 Gödel published his famous proof of undecidability, which demonstrated the limited possibilities of deductive thinking.

It is hardly an exaggeration to state that this was the most powerful epistemological result ever obtained. However, all this is no longer philosophy; it is mathematics. And the foundation of mathematics is not at all a philosophical discipline though it can be traced back to Kant and Leibniz. So mathematics or, to be more exact, some of its sections acquire a philosophical flavor in their formulations of problems.

Analytical philosophy criticized the language of philosophy. Ths led immediately to an interest in the analysis of scientific language. The neopositivists declared that scientific language should be reformed and the natural sciences should be turned into a calculus. When this part of their program failed, a new task appeared: the mathematization of knowledge. But how can we approach the mathematization, say, of biology or medicine if we are vague as to their logical structure?

The philosophy of science has merely turned into a metascience. By this I mean a science which deals with the study of the way certain sciences are constructed: how the hypotheses are stated, how they are accepted and rejected, how the language is constructed in which hypotheses are formulated, how experiments are organized, how inferences are drawn, etc. Today the philosophy of science consists primarily in a rigorous analysis of the means and methods of particular sciences. This analysis is built upon observations and logical constructions subject to verification (or falsification).

As opposed to speculative philosophical constructions, this work elicited a direct response in scientific circles. To verify this, I studied the citation rates, according to Garfield's *Science Citation Index (SCI)*, of some well-known philosophers working in the philosophy of science. Note that citation in this index reflects only the response in journals dealing with concrete investigations in mathematics, physics, chemistry, biology, medicine, and technology, and to only a small degree those dealing with traditional philosophy.[4] *SCI* for 1968 and 1970 was scanned by

---

[4] Recently, the Institute for Scientific Information started publishing *Social Sciences Citation Index*, which reflects citation in the humanities as well. In this *Index*, the list of the most frequently cited authors included the philosophers Dewey, Kuhn, and Popper. Their average citations for 1969–1977 were, respectively, 227, 196, and 171 (*Current Contents*, Vol. 10, No. 38, 1978).

my colleague T. I. Murashova, and the results for the two years are as follows: R. Carnap, 19 and 30; B. Russell, 43 and 48; K. Popper, 24 and 40; L. Wittgenstein, 9 and 11; H. Reichenbach, 26 and 50. The figures are indeed large; no physicist, chemist, or mathematician can boast of such a citation rate in scientific journals. Thus we see that the philosophy of science is a philosophical investigation of the methodology of the sciences. Here concepts from the humanities penetrate the foundation of all scientific activities, including even mathematics.

Physics. It is now possible to speak of the humanization even of physics. The principle of complementarity introduced by Bohr means rejection of one of the principal laws of logic, the law of the excluded middle (Nalimov, 1976*a*), or, in other words, a resort to a metaphor. Thus, one of the extralogical forms of our everyday, nonscientific language has peacefully entered the system of scientific thinking. We know how physics introduced chance and probability into its structure, though European scientific and philosophical thought led an uncompromising struggle with them for more than twenty centuries. Chance stopped being the expression of our ignorance and turned into a way of describing our knowledge. Science, and physics first of all, had to resort to the language of probabilistic concepts (Nalimov, 1976*b*) which had for a long time been used only in our everyday nonscientific verbal behavior. Another thing is worth noting: some physicists (although a small number) have turned to the ideas of ancient philosophical teachings. Let the titles of two works by F. Capra, a well-known American physicist working in the field of elementary particles, serve as an illustration of this: "Modern Physics and Eastern Mysticism" (*The Journal of Transpersonal Psychology*, 8:20–39, 1976) and *The Tao of Physics* (Shambhala, Boulder, Colorado, 1975).

This interest in the ancient Orient is related to the development of the "bootstrap" concept, well known in modern physics. And whereas the penetration of the humanities into mathematics is expressed by the appearance of new problems within this discipline, their penetration into physics is manifest as an enrichment due to a change in its forms of language and to the influence of novel "world views" which hitherto have been alien to mainstream Western science.

Cybernetic Nature of Science

Side by side with the humanization of knowledge, another process is going on: the "cybernetization" of science. Science has acquired new functions: the solving of problems connected with the search for the optimal forms of human activity. This has also increased interest in the humanities and added a humanistic character to scientific activities. At least three reasons can be pointed out that have promoted the cybernetization of scientific knowledge.

1. Engineering branches of technology are anxious to create devices imitating not merely mechanical human activities but also intellectual properties. The problem of man has suddenly come to the fore in the problems of control.

2. The task of controlling the progress of science has led to the need for and development of a metascience, a probing of the foundations of science. The researcher, no matter what his area of research, wants to know whether the methods he uses are legitimate, whether the rules of constructing hypotheses are valid, whether radical changes are necessary, whether the mathematization of knowledge which lately has received much publicity is justified, and what are the foundations of mathematics itself. The unconditional belief in scientific methods has been replaced by a form of critical doubt. Scientific methods have themselves become an object of analysis. A scientist wishes not only to study things but also to control and understand his research practices.

3. The prestige of science has started to decrease. This process is very prominent in the West and has been much discussed in various publications. It is also noticeable in the Soviet Union, as indicated, for example, by the decreased number of applicants to many higher educational institutions, especially to engineering schools.

In the recent past, say, in the time of Pasteur, it was somehow self-evident that science was useful for mankind, even if nobody controlled or directed its progress. At present, this statement becomes more and more dubious. Although nobody claims that the exhaustion of resources, the pollution of the environment, the spreading of some diseases, or the increase of criminality and drug addiction are a direct and unavoidable result of scientific development, the latter has not prevented these phenomena from occurring. And this is indictment enough. It seems that the evolution of science should acquire a goal.

It has suddenly become evident that the consequences of scientific activity, no matter what branch of knowledge the latter embraces or how abstract it is, are directed at mastering nature. Meanwhile, the uncontrolled and arbitrary interference with the global ecology, of which man himself is a part, has proved to be a threat to the future of humanity. The problem has acquired a cosmic significance. This gave rise to the idea that scientific development charged people with the burden of responsibility for which they were not ideologically prepared. In a sense, the whole of science has made man its center. By comprehending this phenomenon we come to see that the whole of science is humanistic.

"Humanization" of Education

Humanization of knowledge affects not only the methods and contents of separate scientific disciplines but also some other processes in science; in particular, this becomes obvious through the changes in higher education. In many countries, science-oriented students are given substantial training in the humanities. For example, most American universities are two-stage institutions. In the first stage, undergraduates attend four-year colleges which provide general higher education and professional training. Graduation from college leads to the first American degree, the Bachelor of Arts or Bachelor of Science. Qualified students are then trained for specialization in various fields of knowledge. (This is a sort of combination of what are called the higher school and the postgraduate school in the USSR.) Completion of the second stage leads to the second American degree, the Master's degree, which approximately corresponds to the university diploma in the USSR. The term of training in higher professional schools is four years for medical and law students, and two years for engineering and economics students.

Now let us consider the undergraduate education system at one of the top ten American schools, Stanford University. There the program for general education is carried on during the four years of studies along with the programs corresponding to the demands of the principal departments the student will specialize in after graduation (a student must declare his future specialty by the beginning of the third year). The program of general education includes English, history, foreign language (including such languages as Chinese, Greek, Hebrew, Latin, and Japanese), and mathematics. The language study must be accompanied by study of the culture and customs of the corresponding country.

All the students must take courses in two fields which are not their specialty: the humanities (eight units), which include fine arts (archeology, art, music, oratory, drama), philosophy, religion, and culture, and the social sciences (ten units), which include anthropology, communication, economics, population geography, political sciences, psychology, and sociology. Additional requirements for the Bachelor of Arts degree include the following disciplines: logic, psychology, and statistics (*Study of Education in Stanford,* vol. 2, *Undergraduate Education,* 1968).

What is the reason for such great concentration on the humanities in the American university? This question is very easy to answer. There is a tendency to broaden the general outlook which is so necessary for understanding interdisciplinary fields of knowledge.

For example, a student studying the resistance of materials can learn how to solve the problems successfully or one studying organic chemistry can sharpen his insight into the methods of chemical synthesis, but neither student will become critical minded. In both of the above-mentioned fields of knowledge, critical consideration of the principal ideas is possible only on the basis of a large background of creative work. Matters stand differently in the humanities. A student studying anthropology or philosophy is exposed to a variety of simultaneously existing hypotheses. Their consideration and comprehension immediately turn into their critical analysis.

In conclusion, this wide humanistic training of the non-humanists is the reflection, perhaps not yet fully understood, of the process of humanization of knowledge, which I dwelt upon above.

Several words should be said about Soviet universities. They have lost their holistic nature, which used to distinguish them from other higher educational institutions in the USSR. The title university no longer suits the essence. At present a university is a mechanical combination of isolated departments. If a certain department turns into an independent institution, neither the students nor the staff of other departments will notice. This has happened with medicine, which was the first to leave our universities and to break the links with other branches of knowledge. Perhaps all this is a result of the differentiation in science which becomes more and more obvious. Even new scientific disciplines emerging at the junction of several branches of knowledge do not long preserve an interdisciplinary status but rather turn into a new, separate subject.

The Need for Staff Educated in the Humanities

My colleagues G. A. Batulova and A. V. Yarkho (Batulova et al., 1975; see also Chapter 12 of this book) studied the changes in the demand for people with higher education in Great Britain during a recent 10-year period (see Chapter 12 for details). The employment advertisements in the *Times* of London showed that the greatest demand was for people with humanistic education: 17.7 percent of the advertisements in 1961 and 25.9 percent in 1971. Economists were in second place (13.7 percent), and mechanical engineers held third place (11.3 percent); each of the remaining specialties comprised no more than 10 percent. It is interesting that the greatest demand for staff with humanistic education came from institutions of higher education. This suggests that the humanization of education has not reached its saturation point as yet. The rest of the demand came from quite a variety of institutions, the main part being private firms and governmental organizations. It appears that modern society lives a complex intellectual life which demands more and more people with a broad intellectual outlook and high general intelligence. Members of this society need a knowledge of foreign languages as well as great skill in the native language, the ability to orient themselves in complex situations, and the ability to absorb novel ideas and make all necessary inquiries to gain data on quite new, previously unknown problems.

Such broad specialists would be welcome in the USSR; all of us would be delighted to have such assistants. However, in the USSR people educated in the humanities generally do not also seek higher education in some narrow engineering discipline. There may be psychological reasons for this: many administrators are ashamed to have on their staff people with a humanistic background, and many engineers, especially women, wish to get more comfortable jobs than those usually occupied by engineers. For example, I. G. Petrovskii, the late rector of Moscow University, in one of his last speeches said that only twenty graduates out of eighty in nuclear physics obtained work in this field. For the rest of them, nuclear physics was no more than general education. In many engineering institutes it is mainly the women who study technological specialties, even though they are rarely engaged as technologists. Why then do they study technological disciplines? We can hardly consider a course in the processes and apparatus of chemical technology as one that widens a student's intellectual outlook.

It seems that adolescents finishing secondary school in the USSR have intuitively comprehended the growing role of the humanities in modern society. There is enormous competition among applicants for positions in the humanities departments of higher educational institutions and a regular decrease in the number of applicants for many other specialities.

Resistance to the "Humanization" of Knowledge

Like any other self-organizing system, science possesses inertia directed at maintaining its stability. It is of interest to see how the structure of science in the USSR resists the penetration of humanistic knowledge.

To my mind, a noteworthy feature of the USSR is the tendency to place the exact sciences and the humanities in sharp opposition. This opposition is built into the system of strict separation of specialists in accordance with their titles: they are divided into the representatives of exact sciences, biological sciences, technological sciences, and the humanities, and this separation is taken quite seriously. In many libraries, including the Lenin library, a Ph.D. in technological sciences may not take out a book on Buddhism: it is forbidden. The Institute of History of Natural

Sciences and Technology of the USSR Academy of Sciences resisted publication of my paper "Predecessors of Cybernetics in Ancient India" merely because I am a doctor of technology. It was an agreeable surprise to learn that Joseph Needham, the author of the well-known five-volume work *Science and Civilization in China*, is not only a sinologist but also a biochemist: on the title page of the book one reads that the author used to lecture on biochemistry at Cambridge University. My book on language (Nalimov, 1974*b*) could be published only because all its constructions were based (though heuristically) on a mathematical model, but at what cost! I have learned that the book is being widely used by institutions dealing with engineering and technological problems. And now, when attempts are made to publish an expanded edition of the book, the only thing which troubles the editing board is the possibility of its having philosophical errors. The question of where the new version is to be published — by the humanities board or by the technological board — has been strenuously discussed for months on end.

Concluding Remarks

What, then, is the "humanization" of knowledge? It is hard to answer this question concisely.

First of all, it means returning to knowledge its lost unity, i.e., its indivisibility. It also means the acknowledgment of the anthropocentric nature of knowledge. We recognize that purely human problems lie behind all other problems. We begin to realize that our entire knowledge is conjugate with man, with specific features of his thinking and his drives, no matter whether they are physical or spiritual. We see that pure logic separated from man and enclosed in the iron box of a computer is only an auxiliary engineering device; in no way is it the source of knowledge.

It is high time to stop looking at the world with photoelements, thermoelements, and other measuring devices and begin to acknowledge our right to look at the world of those who manipulate these devices and interpret the data they produce.

The problem of Man suddenly comes to the fore. Everything is connected with it. We begin to feel acutely the cosmic responsibility of man for the long process of uncontrolled domination over nature. Perhaps the most important thing now is the emergence of the necessity to change radically the whole system of education: make it broader, more humanistic, and, probably, even anthropocentric. This will naturally change into cosmocentricity since it is the destiny of the planet Earth which is at stake.

However, neither humanization nor the emergence of anthropocentric or cybernetic ideas in science solves the problem of Goal, although this problem becomes especially acute and crucial in the expanded and anthropocentric science of the future. But modern logic has taught us the hierarchical language: we know that Goal is a metaconcept, and it should be formulated and discussed in the semantics of the metalanguage — the language of cosmic destinies of the worlds, civilizations, and biospheres, be they real or imagined. But we live on the Earth and all our experience is formulated in a hierarchically lower object language. The only thing which remains to be done is to repeat the words of the Cheshire Cat taken as an epigraph for this chapter:

> *". . . you're sure to [get somewhere]*
> *. . . if you only walk long enough."*

Chapter 10

Is a Scientific Approach to the Eschatological Problem Possible?

A Logical Analysis of the Problem of Global Ecology[1]

For there shall be days when you will say:
Blessed is the womb which has not conceived,
and those breasts which have not given suck.
                    APOCRYPHAL GOSPEL OF ST. THOMAS

Oh man? why is the world becoming so small
for you? You want to possess it alone; but
if you had possessed it, it would not have
been spacious enough for you; . . . ah, this
is the pride of the devil who has fallen from
heaven into hell.
                    JACOB BÖHME

Introduction

In recent years, the ecological problem has acquired apocalyptic over-tones. To solve this problem, science must not only study a phenomenon, but must also learn to predict its evolution on a large time-scale. Further-more, the solution to the ecological problem cannot but change the direc-tion of our cultural progress. Never before has science been faced with problems of such global significance. Is it ready to solve them? To an-swer this question we must understand whether scientific forecasting is possible, whether scientific ideas can influence social behavior, whether

[1] This chapter was translated by A. V. Yarkho.

215

the historical origin of this crisis can be scientifically analyzed, whether a scientific approach to setting a global goal is possible. Below I shall not try to answer these questions but only to discuss them. The problem is so serious that it should be discussed freely and objectively.

My sole aim is to demonstrate that *there may be another approach to the problem, different from the existing one.* This chapter is written in an axiomatic–narrative style. Illustrations and arguments serve to elucidate the axiomatic statements. It is not my purpose to prove anything or to convince the reader. It is up to the reader to regard my approach as legitimate or not on the basis of his own experience and on the facts as he knows them.

Logical Analysis of the Problem of Forecasting[2]

> *Soothsayers.*
> *As lower down my sight descended on them,*
> *Wondrously each one seemed to be distorted,*
> *From chin to the beginning of the chest;*
> *For tow'rds the reins the countenance was turned*
> *And backward it behoved them to advance,*
> *As to look forward had been taken from them.*
> THE DIVINE COMEDY

We have suddenly become aware of the threatening aspect of the ecological crisis as a result of the forecasting of its future development. Therefore, it seems natural to start our discussion with a logical analysis of forecasting.

Formally, forecasting is nothing more than extrapolation. However, we do hope to get precise and definite ideas of the future on the basis of fairly vague notions of the mechanisms which have been operating in the past.

Is scientific forecasting of this type possible? Strictly speaking, the answer is no. In the natural sciences, only those constructions are considered scientific which can be verified by experiment. Here lies the demarcation line between scientific knowledge and non–scientific constructions. This is not "positivism" but the standpoint of a naturalist, which determines his everyday life. All of us are well aware of the difficulties connected with a formal definition of what constitutes the experimental testing of a hypothesis. Verification is logically inept. Falsification, in the terms of Popper, is logically clear but far from universally applicable. One could describe numerous natural scientific construc-

---

[2] This section was published in part in the journal *Znanie-sila* (no. 1, 1972).

Logical Analysis of the Ecological Problem

tions which have not been tested by falsification. The American study mentioned earlier is an ecological illustration. An extensive study of five large ecosystems (the steppe, the tundra, etc.) was carried out, and mathematical models were built that included up to 1,000 parameters. In building the models, experimental data were used, but the model as a whole could not be subjected to direct experimental testing. Thus, as mentioned above in the context of statistical inference and its problems, these models do not inspire confidence (Mitchell et al., 1976). In forecasting, the comparison with reality can be made only at the moment when the prediction comes true. At the time of its formulation, it cannot be tested and, therefore, in its most general form, it has no scientific status. Nonetheless, everybody is making forecasts: industrial firms, think tanks, and the Pentagon, as well as sciencemetrists, economists, and sociologists.

Here I consider the following questions: In what way is forecasting possible? When does it become scientific? The rest of this section is devoted to what I consider to be the facts concerning forecasting.[3]

1. Deterministic forecasting is possible if we are dealing with an isolated phenomenon whose mechanism is known, and if the forecast is made for a period during which the system that encompasses the phenomenon in question will remain stable. Celestial mechanics can serve as an illustration. However, here, too, initial conditions are given with uncertainty, and everything must be recalculated from time to time.

2. A type of forecasting is widely used in engineering. When designing a bridge, an engineer forecasts its strength for decades. He proceeds from his knowledge of metals and their properties, the resistance of materials, and the statistics of constructions, and he assumes that no natural catastrophe will occur for several decades. Even though he knows his calculations are quite accurate, the engineer insures himself by going beyond the data obtained. He adds in a safety factor in a way which would be impossible for an economist or a sociologist.

3. For the contemporary scientific community, forecasting is both possible and impossible. At present we clearly see the revolutionary nature of scientific development (Popper, Kuhn). From time to time, scientific concepts are formulated which have the characteristics of programs of experimental research and predict new effects. In their moment of decline, old concepts are exhausted and hinder further progress. This has been discussed in numerous works (e.g., Barber, 1961; Popper, 1972; Duncan, 1974; Garfield, 1977).[4] But can one speak of forecasting even in moments of revolutionary development? The future is not scooped from the past, as it should be, according to the meaning of the concept of forecasting. There is a carnival of new ideas. One can speak of conjectures, of insight, but in no way of forecasting.

---

[3] I would like to draw the reader's attention to a paper by Taylor (1977) in which the possibility of forecasting social changes is considered with a critical eye.

[4] The last paper gives a dramatic account of scientific forecasts for flying machines. There were favorable predictions by non-scientists, and very unfavorable ones too. One such unfavorable prediction was made by Newcomb, an outstanding mathematician and astronomer, in the days when the first experiment with an apparatus heavier than air was carried out. (The text of Newcomb's article is given as an appendix to Garfield's paper.)

4. From all that has been said above, the absurd nature of forecasts based on expert estimates becomes evident. Which expert should one prefer: a representative of the orthodox majority who support the established paradigm, or a member of the minority who seek new ways? Are there means to distinguish new sprouts from the weeds which always accompany the growth of science?

5. Supporters of forecasting will claim that there are numerous examples of successful forecasts of gross social phenomena. Such examples do in fact exist. One of them, which is quite amazing, is given in the first epigraph to this chapter. Others, almost as amazing as this one, were made by Nostradamus[5] (Leoni, 1961). Scientists sometimes make forecasts of the same power. In what way do they differ from the non-scientific forecasts quoted above? Does the difference lie in the non-scientific character of their arguments? If the forecasts *were* scientific from the standpoint of natural scientists, why did not all scientists accept them, and why did not other scientists learn to forecast on a scientific basis so as to be able to forecast the catastrophic manifestations of our contemporary culture?

6. Marxism, despite its popularity, did not generate mathematical models which would forecast social development unambiguously and in detail. The enthusiasm of American scientists for mathematical models of the economy proved to be ill-founded, as is stated by Leontief (1971), the originator of the trend. And if one wishes to speak seriously of social development, he would have to agree that we can be certain only of its spontaneous nature. In other words, it is impossible to write an algorithm for all aspects of social development, although this is not a conclusion accepted today. I could also speak of the general impossibility of this task, but this would take us too far from the main theme. If the state of matters is such, it would seem natural to resort to probabilistic methods. But the processes under consideration are nonstationary, random ones, and there is no strict mathematical theory for forecasting nonstationary processes. There have have many attempts to solve the problem heuristically (e.g., Box and Jenkins, 1975), but they are unconvincing. Thus, it is possible to speak only of short-term forecasts, not that these do not have immense local importance, especially in Western countries. The situation there is such that the pollution fines are very high, and at the same time, firms try to keep close to the upper bound;

---

[5] Nostradamus (1503–1566) was a physician and astrologer. His "Centuries" are forecasts written in the form of imaginary travels in the future destinies of mankind. Here is one quatrain which came true during World War II: "The Church of God will be persecuted. / And the holy Temples will be plundered. / The child will put his mother out in her shirt, / Arabs will be allied with the Poles." The translator of the Russian edition interpreted the last line in the light of events related to the Six Day War. This interpretation is, though arbitrary, possible, and it seems nothing better can be done in the interpretation of forecasts.

otherwise technology becomes very expensive. So the situation requires forecasting the consequences of spontaneous technical deviations due to the aging of equipment, uncontrollable change in the quality of raw material, change of meteorological conditions, etc.

7. From a purely psychological aspect, each person does constantly forecast some things. He forecasts his future when he takes a new job, gets married, or simply leaves home. We always live in two time scales. Physically, we live in a narrow, point-like interval of time, and mentally, in the sphere of consciousness, we live in the future and act for its sake. This bifurcation of our time scales is a noteworthy phenomenon of our culture; it is the source of inner disharmony. Such forecasts are always probabilistic (Feigenberg, 1972) and essentially personal. It seems pertinent to speak here of the field of initial concepts with a given distribution of probabilities and of a subjective, probabilistically given filter which reflects our system of preferences. To obtain the distribution of weights in the final forecast, one might use Bayes's theorem. The reasoning will be analogous to that in a later section where I discuss the probabilistic model of behavior. By the way, this may explain why the ancient Greeks, sober and rational as they were, consulted the Delphic Oracle and this introduced ambiguity into their decision making. They were well aware of the probabilistic character of predictions (Lifshitz, 1973), and the mysterious and solemn ritual might have stimulated the mechanism of probabilistic predictions by acting as a trigger. Thus, all this may not have been so absurd as it seems now.

8. *Negative* scientific forecasts are indubitably possible. Drawing curves of growth on the basis of the data of the past we may state with certainty what *cannot* happen if the rest of the system remains unchanged, and our statement will be mathematically grounded. About fifteen years ago, I constructed a curve of growth for the staff of a large research institute. The points perfectly fitted the exponent. The extrapolation showed that in 1980 the staff would reach 92,000!

My conclusion, therefore, is that *only negative scientific forecasts are possible, if any such forecasts are possible at all*. What has been said above is not a proof of this statement but merely an exposition.

The idea of an unavoidable ecological crisis is negative forecasting. The book by Meadows et al. (1972), in its negative part, is attractive for precisely this reason. Forecasts concerning the overcoming of the ecological crisis, no matter where they are written, seem to us utterly naive, and not because we do not believe in human creative potential. Any positive forecast presupposes transition to zero or almost zero rate of growth. The necessity for this is obvious. This, however, raises some questions. Is humanity, which lives within the paradigm of our culture, prepared for this radical and decisive step? If this step is made, where will it lead? Isn't there a danger that society will lose its vital potential? Contemporary science is absolutely unprepared to answer these questions. Below I shall try at least to outline the related problems.

Comparative Study of Cultures as a Way to Understanding the Peculiarities of Our Behavior

Now it is natural to formulate the question: What is the source of the long-term process whose future end seems at present so gloomy? Is it the result of some unavoidable human qualities or a manifestation of specific aspects unique to our culture? While we remain within the framework of our culture, our behavior seems quite natural and the only possible one, but acquaintance with other cultures shows that this is not so. Here lies the significance and fascination of philosophical and comparative anthropology.

If we choose the way of comparative historical analysis, it becomes quite clear that the problem of global ecology is a logical fulfillment of the whole European *Weltanschauung*. Christianity followed the tradition whose roots lie in ancient Judaism and opposed man to Nature, directing him to prevail over Nature. In the first book of the Bible we read:

> And God blessed them, and God said unto them, Be fruitful, and multiply, and replenish the earth, and subdue it, and have dominion over the fish of the sea, and over the fowl of the air, and over every living thing that moveth upon the earth.

Hence it followed quite naturally that any society of Christian culture brought up on the Bible regards itself, while formulating its goals, as a formation independent of Nature and, moreover, dominating it. The idea of human superiority over nature deeply penetrated the consciousness of society and opened the way to the unrestrained development of technological civilization. These historical roots are now forgotten, and the idea of human domination over Nature becomes self-evident without regard to how any stratum of society relates to religion.

This becomes more explicit when we consider the traditions of the Indian culture (on the comparison of Indian and European cultures, see Nalimov and Barinova, 1974). The theory of Karma is a concept of a large system, and a human being is but a link in a long chain of events rooted in the remote past and related to the fates of all other living creatures. Here is how this is described by the Czech indologist Ivo Fisher (1969):

> Belief in the circulation of life presupposed the permanently repeated return of any live creature, any individual, to the Earth. A human

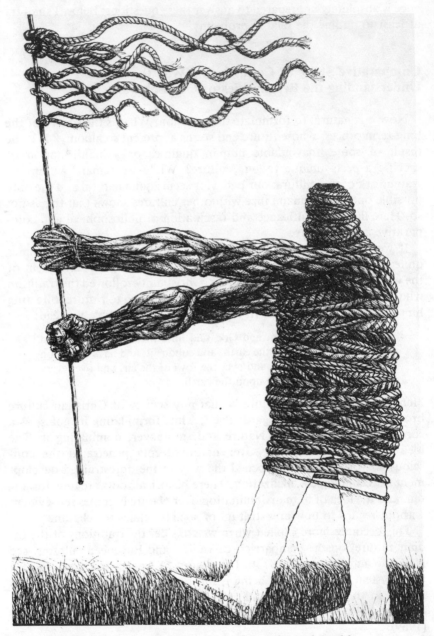

Scientific Approach to the Eschatological Problem

being was thus made equal to all other creatures, and according to some schools, to the world of plants as well. He was only placed at a higher step of development in the framework of the whole process. He was in no way the master of Nature and of living creatures dependent on his will. Reincarnation was regarded as an eternal and immutable law which not only this world obeyed, but also everything living in the Universe. . . . This, of course, greatly influenced the private life of individuals and society, too. . . . Profound comprehension of the beauty of the surrounding world, amazing knowledge of the laws of the animal and vegetable kingdoms, and desire not to harm any manifestation of life were natural consequences of this idea. (p. 51)

We are, of course, far from believing that the solution of the ecological problem will be found by returning to ancient Indian concepts. In their system of views, for instance, certain difficulties connected with the impossibility of placing limitations on the birth rate become especially prominent. Another thing is important for us: the relativity and historical character of those initial premises of our cultures which, even in a cursory survey, seem inevitably human. Our culture goes deep into the past and it is not easy to liberate ourselves from it. Even to claim oneself to be an atheist is not enough.

If the historical analysis is to be continued, it is worth mentioning that Christianity, when fighting the paganism of the Northern peoples of the USSR, came across the belief that the pollution of Nature, the source of the life force for man, was a most profound offense. My father, who was an ethnologist studying the culture of Urgo-Finnish nationalities, often told me about this. He used to emphasize the elevated (toward the whole of nature) ethical tendency of pagan pantheism, which for a long time survived under the cover of Christianity among the people of Komi, which he, himself, came from. In the Western branch of Christianity, the friendly and loving attitude to Nature vividly displayed by Francis of Assisi was at first perceived as a heresy. Later, it entered catholicism, but only as an appendix. Could modern technological civilization have appeared if the prevailing outlook of society was close to the view of Francis of Assisi or of some nature-loving pagans? I do not think it could. The quotation from the Bible cited above opened the way to our scientific–technological progress which has led to the problem of global ecology that we face today. The life of peoples and their ideas endure through ages. For a long time, their sources remain unnoticed, so familiar and unavoidably human have they come to seem.

In conclusion, I would like to mention that G. S. Pomerants has drawn my attention to a paper by the British historian and moralist A. Toynbee (1972), "The Religious Background of the Present Environmental Studies," where the author develops ideas rather close to those set forth above. In his discussion of the ideological sources of the ecological crisis, he, too, ascribes an extremely great significance to the admonition to dominate Nature which the Old Testament gave to European culture. In the conclusion he writes;

> If I am right in my diagnosis of mankind's present-day distress, the remedy lies in reverting from the Weltanschauung of monotheism to that of pantheism, which is older and was once universal.

In my opinion, the overcoming of the ecological crisis requires not a mere transition to almost zero rates of growth, but something more: a radical change in the human attitude toward Nature. This relates to some deep subconscious concepts of our culture. And how can a friendly attitude to nature be combined with limitations of the birth rate?

Dullness as an Index of the Quality of Culture[6]

> *Oh, thou heavy sorrow.*
> *Deadly, wearying*
> *Dullness.*
>
> A. BLOK

If overcoming the ecological crisis requires the creation of a new culture, this immediately suggests the following question: What is the index of the *quality* of a culture? One of the possible answers is that such an index is *dullness*! I am quite aware of the fact this is only a secondary social cateogory. However, I wish to analyze this very category. It is readily grasped, and human consciousness, both individual and collective, responds directly to this index, and not to the deep and obscure phenomena which generate it.

Modern Western society is to some extent protected from boredom by the variety of options available: by mass participation in the creative scientific process, by making careers, by stuffing one's house with a multitude of complicated modern conveniences, and by many other things. This all is carried out as a kind of game, which adds piquancy to life and makes one do his best.

However, this protection is not enough: boredom is coming to the fore, as is manifested by heavy drinking, drug addiction, unmotivated crimes, hippie-like movements that switch part of the young people off

---

[6] A brief version of this section was published in the collection of papers *Value Aspects of Natural Sciences*, Abstracts of the Reports of the Theoretical Conference of the Central Board of Philosophical Seminars of the USSR Academy of Sciences and Obninsk CPSU Town Committee, Obninsk, 1973, p. 79.

the main road of cultural evolution, and by senseless terrorism. Young people naturally respond to dullness most violently.

Almost all "game situations" in modern Western countries require comparatively high rates of growth, so what will happen if growth has to fall almost to zero? The loss of game situations, of pyschic distractions, is always a symptom of the crisis of a culture.

It would be of interest to analyze certain cultures of the remote past from this aspect. We know that there existed viable cultures with zero rates of growth—the so-called primitive cultures (a better name is ritual cultures) where society was protected from dullness by extraordinarily complicated ritual games. Now humanity has grown up, but the threatening question remains: Is it possible to create a culture with zero rates of growth which would protect society from dullness?

Probabilistic Model of Social Behavior

Discussion of the ecological crisis must naturally finish by discussing a model of human social behavior. This model should foresee the possibility of describing the behavior of a person in the unusual conditions which arise when the previous value systems change.

Nowakovskaya (1973) formulates the interesting idea that human behavior can be described in terms of linguistics. In my earlier books (Nalimov, 1974b, 1981), I used the Bayesian theorem to construct a probabilistic model of language which made it possible to explain both the general irregularity of language (word polymorphism) and its logical constituent. Now I am going to show how this model can be used to describe the motivations of human behavior. I do so in the following steps:

1. Human motivations are determined by the value concepts of a person.

2. Value is not a category of formal logic. Logic deals with estimating the truth (or falsity) of statements, but not their value. (In mathematical logic, there does exist a concept of the value of theorems, but this is only an estimation of their complexity.)

3. Something may acquire value if it can be correlated with the concept of a specific goal.

4. A truth, not correlated to anything, has no value.

5. The concept of goal should be formulated in a very concise way, so as to enable the construction of a value scale in the framework of this concept. Statements such as "a striving for survival" or "aspiring for the common good" are too vague and ambiguous to be the grounds for such a goal as, say, changing the human genetic structure. Hence, it is most

unlikely that a large-scale redirection of scientific research, with such a goal in mind, could ever come about.

6. The linking of goal and value is a metaconcept. Here is the formulation of this judgment by Wittgenstein (1963):

> The sense of the world must lie outside the world. In the world everything is as it is and happens as it does happen. In it there is no value — and if there were, it would be of no value (6.41).

7. The goal of the World may be cogently discussed by a metaobserver only if he is in a position to discuss judgments made in the object–language about the behavior of many different worlds or several versions of development of a single world. A metaobserver having at his disposal sufficient experience of the life of various worlds could also give a well-reasoned discussion of the question of changing the human genetic structure.

8. The same difficulty connected with the concept of goal is present in solving local problems. An optimal experimental design is a metastatement which proves possible when the statements about the experiment are formulated in the object–language. The latter, in its turn, becomes possible only after the experiment has been carried out. Here lies the paradox of experimental design — that branch of mathematical statistics which is concerned with choosing the best experimental designs. And it is due to this that the life of a researcher becomes exciting: he plays against Nature when he tries to guess how to choose the experimental strategy before the experimental results are known.

9. Pre-cybernetic science ignored the concept of goal. It was considered theological: a metaobserver is demiurge, the Creator of Worlds.

10. In the cybernetic system of ideas, a live organism is a self-organizing adaptational system capable of creating continuously (or sufficiently often) changing goals in the process of adaptation to the continuously changing world. These are metajudgments of an uncertain character; they are based on an insufficiently rich set of objective statements.

11. From the standpoint of a metaobserver, the development of an adaptation system will be regarded as a walk on a multi-extremal surface. As in the case of numerical methods for locating extrema, the system may become trapped in a logical ravine and perish. We know from history that some cultures, e.g., ancient Rome, perished because they were unable to change the direction of their development; other cultures, e.g., Japan and the Moslem East, managed to adapt themselves and survive under conditions outwardly quite alien to their initial conceptions.

12. A situation of the "Great Inquisitor" type when the goal is rigidly

set once and forever cannot serve either because this is not an adaptational structure but a demiurge-like one.

13. The development of society, even when influenced by science, may
be viewed only as an adaptational walk along some local precipice since a
rigidly given long-term goal does not and cannot exist.

14. Faith in a goal may spread in society like an infectious disease.
The well-known mathematical models of epidemics may be used to
describe this phenomenon. It is quite senseless to look for formally
logical reasons for the appearance of a goal. At the moment of historical
cataclysms, when masses of people rush forward to novel goals, the latter
seem evident to all participants. Retrospectively, however, they often
arouse our amazement, e.g., the Medieval Crusades. It seems possible
that the generations to come after us will be surprised by the branch of
modern culture where consumption is dominant and local optimization
of everything turns into a goal; this is a case of means turning into a goal.

15. Goals may also spread epidemic-like in quiet periods of history.
Let us consider this in somewhat greater detail. In the USSR, despite all
protective measures directed against harmful ideas, complete success has
not been achieved: smoking by young women, guitars, and the long hair
of young men are hardly advocated by anybody, either explicitly or implicitly. They just happen. Of a similar spontaneous nature are the
oscillations in the number of applicants to higher educational institutions: without any obvious reason, the number of applicants to schools
of science and engineering decreases while that of applicants to
humanistic faculties increases. If we now turn our attention to the West,
we shall see that the movement of hippies—or, in the terms of Reich
(1974), of the "third consciousness" of America—can well be described
as an epidemic phenomenon. This movement even does without verbal
argumentation: a fight with the Word, i.e., a contempt toward language,
is one of the essential components of this ideology. Even the intellectual
field created by science and scientific enlightenment does not hamper the
spreading of irrational ideologies. As an illustration, consider fascism.
Even scientists were seized by delirious ideas and in deadly earnest investigated such absurd problems as a military–technological application
of the idea of a "concave Universe."

16. Now let us consider in detail what traits characterize the spreading
of a disease epidemic: suddenness of occurrence; possibility of a latent
state and awakening after a change in the external conditions (in the case
of bacterial infections this may simply be a change of meteorological
conditions); an incubation period without noticeable manifestations of
outward symptoms and its passage to an active state, resulting in damage
to vital centers of the organism or of society; a branching character of

spreading by way of transferring infection from one individual to another; self-exhaustion and decrease to zero values; creation of increased mutageneity, the means of resistance being present. One can notice that all these traits are characteristic not only for viral or bacterial infections but also for ideological ones. The peculiarity of socio-ideological infections is that they prove to be a specific response to stability or stagnation: society is afraid of boredom (as discussed above), and this fright represents a genetically built-in mechanism of its development. Here, probably, an analogy with immunity and the production of protective antibodies is pertinent; in society there are people, carriers of anti-ideas, who oppose the old ideas without generating new ones. Just as the creation of immunity necessitates the appearance of mutations in bacteria and viruses, the weariness from previous ideas gives rise to new directions of thought.

Many ideas may seem new, but after a thorough analysis they prove to be genetically related to ancient ones, and so the history of cultures might be regarded as paleontology of ideas. And if the appearance of new ideas as a response to boredom is considered a natural process, the impact of science on the social process should be directed not at the maintenance of social stability but rather at creating conditions favorable for generating wise changeability.

17. If a goal has arisen, the individual has to make decisions about actions on the basis of the system of postulates given by his value concepts. Here everything is complicated by the fact that postulates in the system of value concepts have different weights which do not remain constant in time. The process of decision making itself may be presented as a two-staged procedure. At the first stage one has to make a decision concerning the reconstruction of the system of postulates; old previously formed concepts (prejudices) must change under the effect of the ideas arising in deliberating the new goal, and this should be accompanied by renormalization, i.e., changing the weight of the postulates. At the second stage logical judgments are based upon a refined system of postulates, and the final decision or strategy of behavior is formulated.

In describing the former procedure, a Bayesian model is quite pertinent. As already mentioned, I earlier used the model to describe the way a man perceives a word on the basis of a field of its meanings. Separate fragments of the field are associated with the symbol–word in human consciousness with various probabilities. This probability is given by the prior distribution function of the word content. In the process of reading a particular concrete phrase, a new distribution function of the same word content is being built, already conditioned by the phrase; then, in accordance with the Bayesian theorem, a posterior distribution function is obtained that takes into account both the past and the new human experience. The essential feature here is re-normalization: what has been placed in the tail part of the prior distribution function may acquire a large weight after the new text is read.

Here again is the Bayesian theorem:

$$p(\mu|y) = kp(\mu)p(y|\mu)$$

When discussing the problem of value, we may ascribe the following meaning to the distribution function: $p(\mu)$ is a priori given (on a segment) distribution function of prejudices in achieving goals of type μ (these may be said to be weights of separate fragments of the value field appearing in achieving goals of type μ); $p(y|\mu)$ is a distribution function for value concepts that arise in deliberating the given concrete problem y, related to problems of type μ; $p(\mu|y)$ is a posterior distribution of values in the situation corresponding to the solution of the given problem; and k is a normalizing multiplier. In a continuous model, squares under the curves given by the functions in the right and left parts of the inequality must equal unity. The posterior distribution function turns into a prior one in solving the next problem of the same type. Below I illustrate the possibility of applying the Bayesian theorem with a concrete example, which needs certain preliminary explanations. It deals with the opposition of society and scientists themselves against new ideas. (I already spoke of this briefly in Chapter 1.)

The history of science is abundant with vivid examples of hindrances which became rather tragic as soon as they got the support of the state or such powerful institutions as the church. One can look back to Anaxagoras, the first materialist–physicist, who was accused of impiety; Socrates, who was put to death by an ignorant Athenian hoi polloi; Giordano Bruno, who, at the beginning of the Scientific Revolution, was burned at the stake by the Inquisition for advocating the plurality of Worlds; Galileo, humbled by the Inquisition; Michael Servetus, burned by Calvin in Geneva, for advocating the circulation of the blood; the Nazi attacks against the theories of "Jewish" physicists, Einstein et al., led by two Nobel Laureates, P. Lenard and J. Stark; Lysenko, whose primitive genetics had the support of the state but was none the less false.

Such situations were not always a result of ignorance or prejudice. Garfield (1977) in his very interesting article "Negative Science and 'The Outlook of the Flying Machine' " described the hindrances before the first flight as follows:

> On December 17, 1903, at 10:35 a.m., Orville Wright took off at the controls of "Flyer 1," flew for 12 seconds, and landed safely—the first controlled, man-carrying mechanical-powered flight in history. But almost five years went by before it was generally accepted that

the Wright brothers had flown in their machine. After all, who were the Wright brothers to make such a claim when the most learned professors, including Professor Simon Newcomb—had "proved" that powered flight was impossible? (p. 8)

The name of Newcomb was well known: he was professor of mathematics and astronomy at Johns Hopkins University, a founder and first president of the American Astronomical Society, and vice-president of the National Academy of Sciences. He also directed the American Nautical Almanac Office. As a scientist, he has still not lost his significance up to now. According to *Science Citation Index*, in the 16-year period from 1961 to 1976, Simon Newcomb was cited 183 times.

Prejudice, armed with logic, is apt to be more harmful to the development of science and philosophy than mere ignorance and superstition. A classic example of the latter was the rejection of the mathematical papers of Evariste Galois by the most eminent mathematicians of France. Group theory was an idea before its time, as was non-Euclidean geometry. Note the refusal of Gauss to publish on this subject for fear of ridicule.

In what way is it possible to explain this stubborn resistance to anything new or unfamiliar, which is catching even for scientists? A general answer to this question is easy. At the early stage of its development, mankind passed through the epoch of magical culture whose one manifestation was unconditional stability guarded by a set of rigid taboos of a sacral nature. [Magical culture preserved in its original form is described by Horton (1975).]

Like Jung (1965), I believe that our unconscious contains everything we have lived through in the process of ethnogenesis, including here the sacral fear of the new. The taboo of magical culture is inherent to the depths of our consciousness just as the urge for revolt is also inherent in some people (see Chapter 2 of this book).

We have only to trace the way in which this old feature from the storehouse of the unconscious manifests itself in our creative life, including scientific activities. As an object of study, I shall take negative reviews and as a model I shall apply the same Bayesian theorem: everything new is filtered by personal perception which stands on guard against it. Now let us turn to our illustration.

The manuscript of my book *Probabilistic Model of Language*[7] (Nalimov, 1974*b*), in which the probabilistic approach to language semantics briefly described above is dealt with in detail, was sent to one philosopher for a review. The manuscript was not strictly philosophical,

---

[7] Published in English under the title *In the Labyrinths of Language: A Mathematician's Journey* (Nalimov, 1981).

but the editorial staff, to be on the safe side, still decided to check on the presence of philosophical errors and for this purpose chose a philosopher with a special attitude. His review affords us a clear picture of his personal prior distribution function $p(\mu)$ giving those requirements which, in his opinion, a good philosophical paper should meet (here μ is a field of values for evaluating philosophical works). His system of value postulates seems to be so constructed that such a book should be thought suitable if much attention in it is paid to opposing the harmful philosophies of the capitalist West: Machism, religion, or to the manifestations of class struggles in modern times, e.g., the struggle against nationalism. Further, it is possible to imagine a function $p(y|\mu)$ constructed while reading the book in question. The reviewer could not but notice that its principal subject was the probabilistic model of language, its interpretation, the semantic scale of languages, etc. Then he could see that I touched slightly upon the questions usually discussed in criticizing Machism and that I was interested in the languages of religion and, though very briefly, mentioned the specific features of ancient Hebrew books. Finally, it is possible to imagine the way the posterior distribution function $p(\mu|y)$ looks, i.e., the way the scale of estimates will look with respect to evaluating the given concrete problem y. Everything connected with the main content of the book is simply lost here since the prior distribution function is arranged so that to these statements there is ascribed precisely zero value. The posterior distribution function will be such that, from its interpretation, it will follow that in the review much attention must be paid to the way the author considers Machism, religion, and struggle against nationalism. Proceeding from these value concepts, it is not difficult to pass to the second stage: constructing a system of clear logical judgments concerning the merits and demerits of the manuscript. All this results in a review saying that the manuscript is of absolutely no scientific value: "the author had at his disposal materials he could use to fight against Machism, but he has not succeeded in it . . . he resorts too often to religious texts, which is inadmissible for a Soviet scholar . . . the author has not drawn a due conclusion in favor of the Russian language and scientific outlook, which is especially objectionable when all honest-minded people, including working Jews, are fighting against Zionism . . . Indeed, mentioning in passing Hebrew books written only with consonants, the author does not use this to fight against criminal Zionist activities."

All these remarks are directed toward a book which in no way pursues political or propaganda goals, but is purely scientific. Its subtitle is *On Relation of Natural and Artificial Languages*. In the book I developed my ideas concerning language with a quite clear-cut task in mind: I wished to understand in what way scientific language should relate to everyday language, in what way scientific terminology should be built, what is the role of mathematics as a language used to describe the external world, what is the implication of the fuzzy semantics of everyday words, and in what way people understand each other when using such words in their speech. Without being able to answer such questions, we cannot solve the problem of artificial intelligence, or that of a man–computer dialogue. To be ready to solve engineering problems, we have to make use of all information on human language, including that concerning cultures of the past. But the reviewer turned out to be a philosopher who perceived the text through a very narrow filter. Everything said above had no interest for him. Despite his quite deprecative estimation of the book, it was published by the publishing house Nauka without any essential corrections, and reviews of the book in both Soviet philosophical journals, *Problems of Philosophy (Voprosy Filosofii)* and *Philosophical Sciences (Filosofskie Nauki)*, were favorable. Philosophers and linguists with other initial value concepts had no difficulty understanding the book.

This example has the flavor of an anecdote, but it is taken from life. Moreover, the editor-in-chief took the review quite seriously and ordered that the manuscript not be sent to the printing house. The example is interesting because it shows how it is possible to make a very strange evaluation by means of logically precise methods: everything can be explained by the fact that a narrowly built system of postulates was used to solve a much more broadly formulated problem. I believe that such cases of odd decision taking are widespread.

18. The probabilistic model must not be regarded too straightforwardly. Of course, I do not think that in reality a person constructs the above-mentioned distribution functions and then multiplies and renorms them. This is just an attempt to give a formal description of the complicated socio-psychological process of whose mechanism we are, strictly speaking, ignorant. It is an attempt to a view a complex system in a certain aspect; this is just a metaphor — in reality, a person behaves "as if but not so" as in the model. The meaning of the model lies in that it allows us to imagine how, in the process of constructing value judgments, the postulates are formed on which further logical judgments are based. The crucial role here is played by the previous background of a person, which determines his system of prejudices. But prejudices manifest themselves in a familiar and, consequently, clear way while the problems previously faced are being evaluated. It seems reasonable to believe that, in solving new problems, prejudices are transformed as suggested by the Bayesian model, and in this situation such odd cases may arise (as in the above example) that they are readily perceived as irrational. I believe this is the kind of behavior to be expected in discussing the problems of global

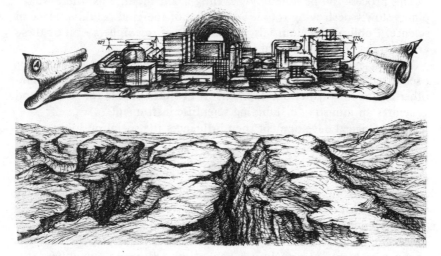

Triumph of the Ecological Problem

ecology, especially when its eschatological character becomes universally acknowledged.

The Role of Science in the Evolving Ecological Crisis[8]

It may be asked what has caused the dominating role of science, scientific activities, and scientific outlook in our society, and the concurrent shift into the background of all other manifestations of human spiritual life.

A simple answer emerges if the development of science is juxtaposed to that of biological species. In biology, separate species, at least at the initial stages of their development, increase numerically according to the law approximately given by an exponential curve. Then, when external resources are running short, the exponential curve turns into a logistic or some other S-shaped curve with satiation. Observations show that a biological system is structured so that it transforms the Earth in a manner favorable for itself, improving its ecological background.

---

[8] A brief version of this section was published in the collection of papers *Value Aspects of Natural Sciences*, Abstracts of the Reports of the Theoretical Conference of the Central Board of Philosophical Seminars of the USSR Academy of Sciences and Obninsk CPSU Town Committee, Obninsk, 1973, p. 79. An adapted version was published for a wider audience in the journal *Izobretatel' i Ratsionalizator* (No. 7, 1977).

This process can be noted by traveling in the mountains: there we see plants slowly destroying rocks. Something of the kind is taking place in science. Publications, journals, the number of researchers—all of these grow exponentially, provided the resources are sufficient.

Thus, science as a system creates ecologically favorable conditions for itself. Under its impact, technology develops, which provides funds and other means for further development. It liberates people for science and generates an industry for building scientific instruments.

Technological development then influences the social outlook: a new scale of values emerges. The old prestige values—heroism and courage, independence and pride, poetry and dreaminess, sympathy and mercy—sink into the background. Science in its progress transforms the surroundings favorably for itself. It is now approaching satiation, which will destroy it before the biosphere is destroyed.

If the number of people engaged in science and the funds allotted for its development do not grow exponentially and, therefore, the scientific community starts aging without being replaced by new members (the signs of which we can already see), and if new technological means necessary to carry out a permanent intensification of experiments cease to be introduced, then the stagnation of science will be unavoidable.

At the moment we sense that the prestige of science is decreasing. At least the prestige of the exact and engineering sciences is declining, as evidenced by the decreasing numbers of applicants to higher educational institutions of this kind in the USSR. People in science have become aware, more intuitively than consciously, that something is going wrong with science, even that something terrible is happening to it. And science is aquiring humanistic tendencies, which are manifested in two ways. On the one hand, representatives of exact sciences have begun to show an interest in basic human problems—language and thinking, psychology, anthropology. On the other hand, science is more and more directing its attention to the universal human problems: the studies of oncology, psychiatry, global ecology. In a number of countries, science is being asked to tackle the problems of controlling social development and forecasting its course.

Concluding Remarks

Thus we see that human social behavior, despite our ability to think rationally, is still rather irrational: ideas regulating social behavior are spreading like an epidemic. The concept of social goals cannot be rationally grounded. Forecasts on a large time scale, if possible at all, can only be negative; progress seems to be nothing else than a walk on a multiextremal surface—on the rough ground crossed by ravines. Probably this irrationalism breaking through the rationalism of our thinking is just what makes life interesting and meaningful. Otherwise it would have been possible to foresee and program everything, and we would only have to fullfill it pedantically and patiently.

The uncertain character of social behavior creates a game situation. Scientific criticism, common sense, experience accumulated through the ages, logical analysis of situations, and ethical concepts which seem to be built in genetically may all be merely elements of a cosmic game!

But who is our partner in the game? Is a game with only one player possible? A model of such a game is solitaire. Shuffling the cards, the player himself generates the situation of randomness against which he is playing. The rules of the game are such that extra cards are discarded, but if the game is not won, the discarded cards are returned to the pack, and it is shuffled again. Does not the same thing happen in the history of mankind? At moments of crisis, old, completely forgotten ideas come to the surface, and so the pack of cards is shuffled again.

The ecological crisis is primarily an ideological crisis. It is the crisis of Western culture and its concepts of goal and value. Contemporary culture needs respiritualization. Nobody knows from whence new ideas will appear. We come to understand the new interest in the half-forgotten ideas of past epochs.

The role of strict scientific thinking in solving the problems of global ecology is that of providing an acute critical analysis of these problems and of the spontaneously arising methods for their solution. Critical analysis, if sober and bold, may stimulate human creativity before the crisis becomes irreversibly sinister.

Social development, even when influenced by science, may be regarded only as an adaptational walk on a surface crossed by local ravines since no long-term goal does or can exist. The change in the cultural tendencies is, after all, no more than changing the initial point on a multiextremal surface whence the movement begins, as well as changing the direction of the gradient, since the goal function is changed.

Chapter 11

Geographic Distribution of Scientific Information[1]

Introduction

This chapter is aimed at drawing attention to the possibility of elaborating a new trend in sciencemetrics: namely, the study of the distribution of scientific journals among the various countries of the world and among scientific centers within individual countries. This problem might be called an ecological problem of the development of science. The expenditures on science made by each country are limited as compared with the needs of scientists. According to Price (1967), each country spent 0.7 percent of its gross output on the development of science.[2] It remains unknown what part of these expenditures goes for obtaining scientific information by a particular country. It is also not clear how these expenditures are divided among scientific centers within a given country. We know that the gross joint output of economically advanced countries doubled in about 25 years whereas the number of scientific journals doubled in about 12 to 15 years. Thus, the difficulties of obtaining the whole variety of journals published must increase with time. This fact, in turn, will result in greater unevenness in the distribution of scientific journals, to the detriment of those who have not enough of them as it is. Here we see again a manifestation of the Matthew effect

---

[1] This chapter was written with I. V. Kordon and A. Ya. Korneeva and was published in *Information-nie Materiali (Informational Papers)*, published by the Scientific Council on Cybernetics under the Soviet Academy of Science (Moscow, 1971). It was translated by L. R. Moshinskaya.

[2] These data differ considerably from the data of the official statistics which, naturally, cannot clearly separate the expenses for purely scientific investigations and the expenses for technical and military investigations. Price's calculations are based on records of the number of publications, their cost, etc.; i.e., he has made an attempt to estimate the actual scientific expenses by using indirect data.

(Ziman, 1969): whosoever hath, to him shall be given, and whosoever hath not, from him shall be taken away. To some degree, this effect aggravates preexisting rivalries. In any case, one can be certain that the increased cost of scientific information will cause an increase in the unevenness of the distribution of information among scientific centers. Not only scientists of the various countries of the world but also the scientists at various scientific centers within the same country may find themselves in unequal conditions. And this will inevitably have an effect on the development of science proper: scientists with the different degrees of information will have essentially different scientific potentials.

In this first treatment of a quite novel topic, we cannot give an in-depth account of the above-stated problem. However, we do offer three examples which suggest that investigations of this problem can yield important results.

The Stocking of the Library of Odessa University with Scientific Journals

We begin this section with a thorough analysis of the process of stocking the Odessa University library with scientific periodicals from the beginning of the nineteenth century up to now. This illustration shows quite vividly the initiation of the process of hampering the development of science (changing the exponential growth into logistic growth) which was so heatedly discussed by sciencemetrists, of whom the first was D. Price. Organizationally, the development of science can be regarded as the evolution of an informational process (Nalimov and Mul'chenko, 1969). It is of interest that the stocking of the library, which is essential to this process, proved to be just the element in the complex structure that was the first to be affected by the events that hampered the growth of science.

The history of the Odessa University scientific library began in 1817 when the Richelieu Lyceum was founded. Richelieu, the governor-general of Odessa, presented it with his private library. The Richelieu Lyceum was an educational institution without a specialization. In 1837 its structure was changed so as to make it closer to a higher educational institution, and in 1865 it was rearranged as Novorosiisk University. It later underwent many reorganizations, growing from 175 students in 1865 to 12,000 in 1961.

To trace the growth of the Odessa University library, we determined the number of journals it acquired in various fields of science over a period of 150 years. We made an attempt to approximate the growth curves by exponential and logistic curves. Figure 1 shows the graphs obtained for the field of geography and is representative of graphs obtained also for the natural sciences, mathematics, technology, geology, economics, chemistry, and physics.

The approximations did not exactly reflect the actual process in detail. This should be expected because this university, in the process of its development, had undergone numerous changes which influenced the availability of library funds, but these details are not of great significance for our task. Nevertheless, the approximation of the growth curves by the exponential and logistic curves allowed us to fix the moment when the mechanism of growth changed. Knowing the parameters of logistic curves (they have been calculated for the whole set of the values observed), we can easily find the abscissa value corresponding to the points of inflection where the sign of the second derivative (the sign of acceleration) changes. It turned out that on almost all the graphs the point of inflection of the logistic curves occurs in the region of 1900. However, as can be seen in Fig. 2, for biology the point of inflection is quite different — it is elevated as a result of the poor approximation of the logistic dependent data.

The same results were obtained in the approximation by exponentials. In calculating the exponent parameters, we took 1910 as the last point on one occasion and 1890 as the last point on another.[3] In the second case the approximation was much better; this also indicates that the mechanism of growth was changing somewhat about 1900. So we see that during the nineteenth century the stocking of the library by journals was growing exponentially. The exponent parameters (calculated with the last point in 1910) are such that the doubling of the number of journals was as follows: geography during 16.8 years; natural science, 17.2 years; biology, 15.0 years; economics, 13.2 years; chemistry, 22.5 years; and physics, 17.8 years. These rates of growth correspond approximately to the rate of growth of the number of journals in the world at that period. At the boundary of the two centuries, financial difficulties appeared; they manifested themselves particularly clearly after 1910. The rate of growth changed from exponential to logistic. This condition persisted until 1950, when the USSR started producing photocopies of some foreign journals, but this practice ceased after the USSR signed the International Copyright Treaty. At the same time, the number of students at

---

[3] When describing the growth curves in sciencemetrics, one has to use the sliding-exponent model whose parameters remain constant only in a comparatively small time interval; the first estimation showed that good approximation can be expected only for an interval of about 20 years. It is also noteworthy that so far in sciencemetrics the logistic model of growth has been considered only as logically possible; here for the first time we have obtained the real data which in part are quite sufficiently described by the logistic dependence.

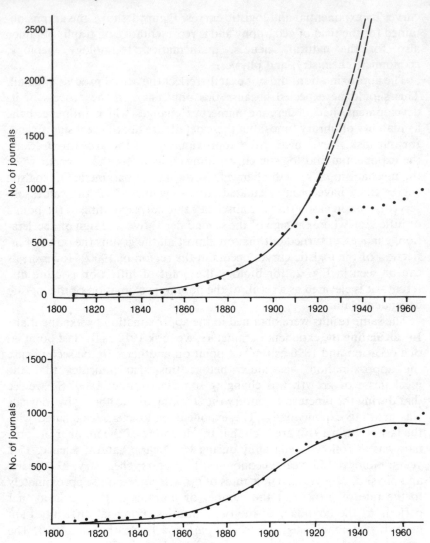

FIGURE 1.   Growth of the number of journals acquired by Odessa University Library in the field of geography. Top part contains the observed (dependent) data (•); the curves are exponential approximations fitted to the data. Lower part contains dependent data and a logistic curve fitted as an approximation.

the university began growing substantially after 1895 and by 1970 was 20 times the number in 1900.

Here we again remind the reader that the growth of scientific periodicals still takes place according to a sliding exponent (with the

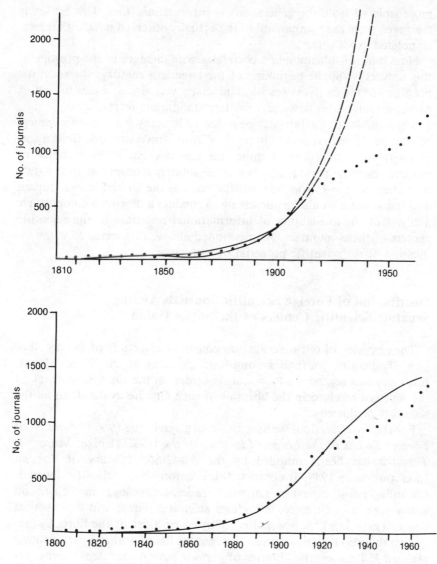

FIGURE 2.   Growth of the number of journals acquired by Odessa University Library in the field of biology. Top part contains the observed (dependent) data (•); the curves are exponential approximations fitted to the data. Lower part contains dependent data and a logistic curve fitted as an approximation.

disturbances during the two World Wars), and the exponent parameters are so changed that the number of journals doubles during a shorter period. Gradually, the Odessa University library has become more and more isolated from the general world information flow. This is clearly seen even from the comparison of the actually observed data with the extrapolated exponents.[4]

Note that this phenomenon correlates with changes in the prestige of this university: at the beginning of the twentieth century, the scientific prestige of Odessa (Novorosiisk) University was almost equal to that of Moscow University; nowadays matters stand quite differently.

The same state of affairs can be observed in many old scientific centers of Europe. If one visits the library of Vilnius University, one finds a surprisingly rich collection of quite old journals but a poor choice of modern scientific periodicals. We are absolutely ignorant as to the state of matters in non-European countries such as the United States, Japan, and India, but it would be interesting to conduct a thorough comparative analysis of the availability of informational resources in the scientific centers of these countries. As mentioned above, this seems to us an indication of the scientific potential of a country.

Distribution of Foreign Scientific Journals Among Separate Scientific Centers of the Soviet Union

The next step of our investigation consisted in a study of the distribution of scientific journals among separate cities of the Soviet Union. Each city was regarded as a scientific center on the supposition that all the journals received in the libraries of each city are available to all the scientists of this city.

For this investigation, we used the data given in the 1966 *Union List of Foreign Serials in the Largest Libraries of the USSR* (Kniga, Moscow). This list has been compiled by the All-Union Library of Foreign Literature since 1949. It contains information about scientific journals (including photocopies) on natural sciences, technology, medicine, and agriculture, as well as some popular scientific journals in alphabetical order. The title of the journal is followed by a list of the libraries (encoded) that receive this journal. In 1966, the list of libraries participating reached 354 names, the libraries of various government departments being excluded. To give an impression of the libraries scanned, we note that the list includes 155 Moscow libraries, 8 Odessa libraries, etc. In any case, the catalogue scans all the libraries available for a wide range of

---

[4] One might object that the publishing of abstract journals in the USSR since 1953 made up for the deficiency of the original journals, but abroad the abstract journals appeared much earlier and caused no decrease in the circulation of primary journals. Nobody can convince a scientist that subscription to the abstract journal, even with the possibility of ordering photocopies of the publications needed, is equal to obtaining the primary journals. If it were, there would be no need to publish primary journals: it would be much easier to publish only abstracts and to deposit the manuscripts.

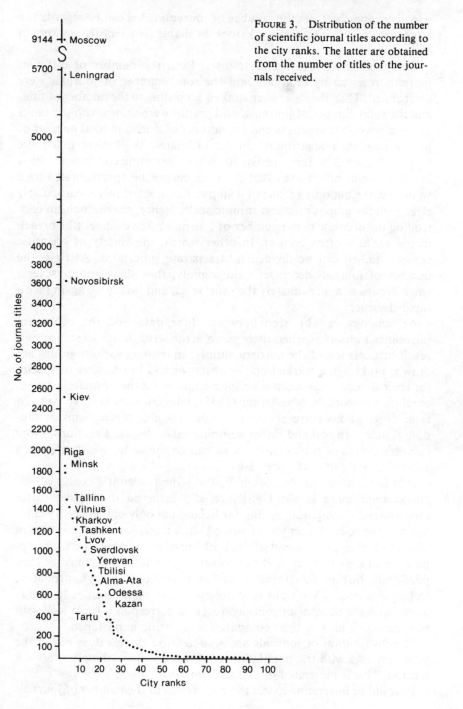

FIGURE 3. Distribution of the number of scientific journal titles according to the city ranks. The latter are obtained from the number of titles of the journals received.

scientists, and therefore the number of journals listed can be regarded as a measure of the informational stock available to scientific workers in each center.

The data were processed as follows. Both the number of different journals received by each city and the total number of journals were determined. Then the cities were ranked according to the number of titles and the total number of journals, and graphs were drawn with the ranks of the cities on the abscissas and the number of titles and total number of journals (or their logarithms) on the ordinates. We believe that these graphs (Figures 3–6) form a basis for some interesting conclusions. First of all, it seems noteworthy that the data cannot be approximated by a smooth curve but only by one of a stepwise character: this is particularly clear with the graphs on a logarithmic scale. Hence, the mechanism controlling the decrease in the number of journals often changes (the breach in the variable often occurs). In other words, the variety of scientific centers studied can be divided into separate subgroups. At first, the number of journals decreases quite rapidly; then there appear several small groups at approximately the same level, and finally we again face a rapid decrease.

An analogy can be seen between these data and the ecological phenomena observed when there is an acute struggle for existence as a result of depletion of the nutrient supply. In ecology various graphs are drawn, ranks being marked on the abscissas and the number of species (or their biomass) for a certain region or basin on the ordinates. An example is the work by MacArthur (1957), who considered three types of rank distributions corresponding to three possible mechanisms of random ranking. In real and rather complicated ecological situations, more complex curves of rank distributions can be observed, which are often similar to our curves (Figures 3–6).

It is interesting that Novosibirsk, the largest scientific center of all-Union importance in the USSR, rapidly achieved third place (after Moscow and Leningrad), leaving far behind not only old university cities but also the scientific centers of some Union republics. Odessa is now in nineteenth place. At the end of the rank correlation curve, there also appear the university cities, and the university, by definition, must be supplied with information in all fields of knowledge, except, probably, technical sciences. We want to emphasize that ranking cities in accordance with the number of journals received corresponds quite well with our subjective idea of their comparative scientific importance.

The distribution of journals among scientific centers does not in the least correlate with the population of the cities where these centers are located. This is illustrated in Figures 7 and 8.

It would be interesting to test the correlation of the number of journals

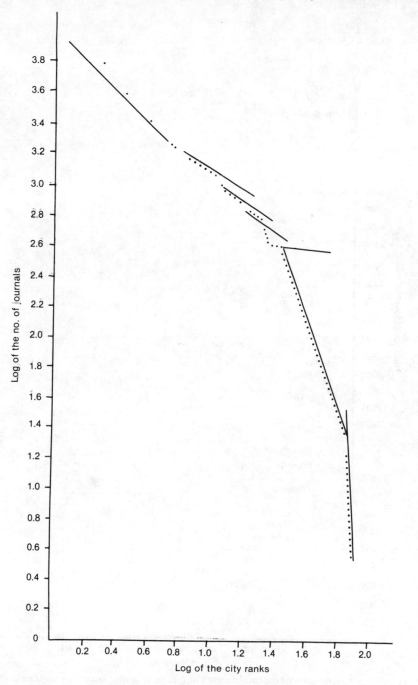

FIGURE 4.   Data of Figure 3 after logarithmic transformation of the scales.

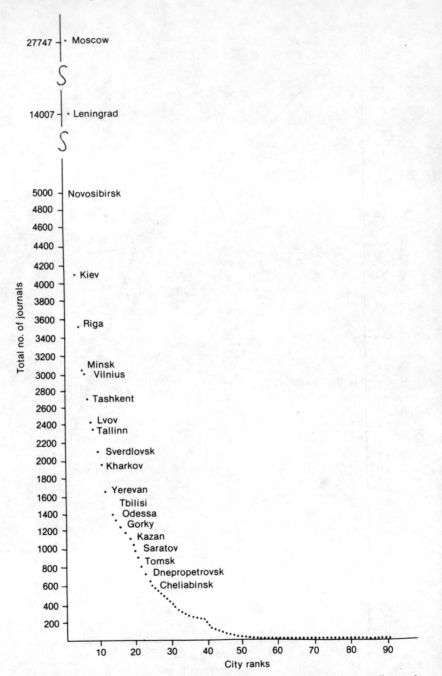

FIGURE 5.   Distribution of the total number of foreign scientific journals according to the city ranks. The latter are obtained from the total number of journals received.

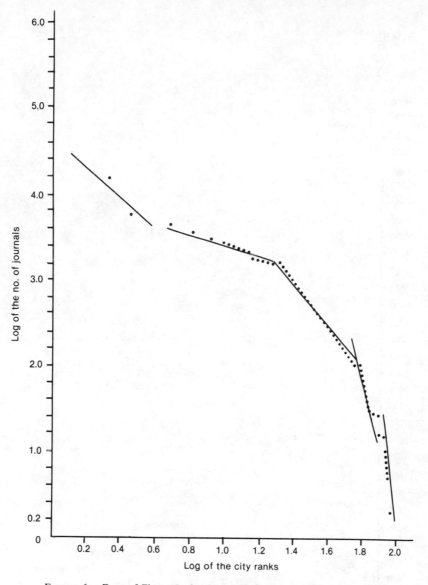

FIGURE 6.   Data of Figure 5 after logarithmic transformation of the scales.

with the number of scientists, but unfortunately we do not have the necessary data at our disposal. We have available only data on the relative distribution of creative scientists in the USSR among the four most important (in terms of scientific productivity) centers. The data we use here were published in a paper by Price (1970), who had obtained

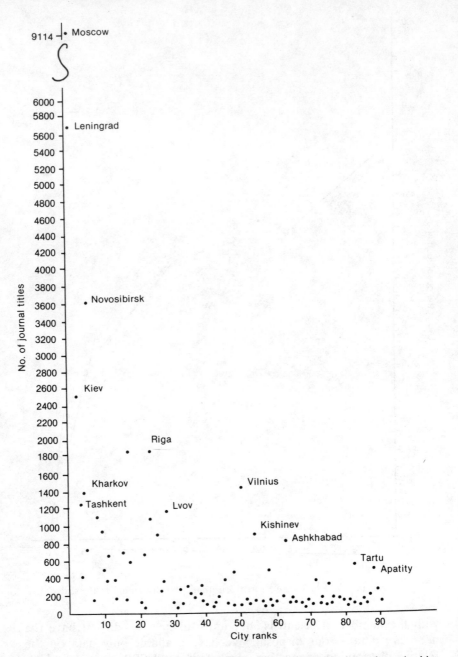

FIGURE 7.   Comparison of the number of titles of foreign scientific journals received by the cities with their ranks according to population.

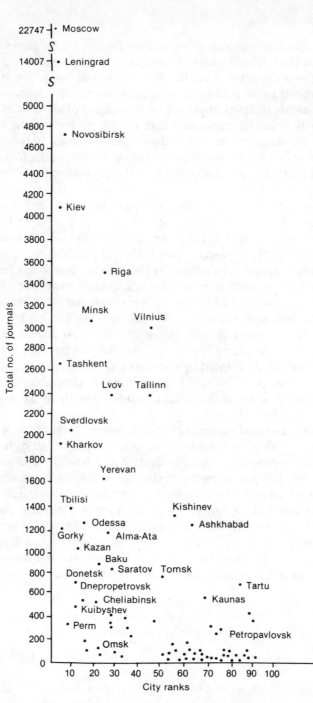

FIGURE 8. Comparison of the total number of foreign journals received by the cities with their ranks according to population.

249

them from the *International Directory of Research and Development Scientists* for 1967. This directory is compiled by Eugene Garfield on the basis of *Current Contents*, a publication which contains information on scientific articles to be published. A special test has demonstrated that *Current Contents* includes about 80 or 90 percent of all articles which turn out to have considerable influence (measured by number of citations) upon the development of world information flows. In the *International Directory of Research and Development Scientists*, the first author of each publication is listed according to the country, city, and institution to which he belongs.

In comparing the data on the distribution of creative scientists and the distribution of foreign journals in four cities in the USSR, we obtained the results shown in Table 1. It can be seen that, though the concentration of journals in Moscow seems very high, it lags behind the concentration of creative scientists. This suggests that we are witnessing a feedback process: the high concentration of journals, created under the influence of the high concentration of scientists, has stimulated a further increase of the concentration of scientists. This spontaneous process encounters certain resistance on the part of the authorities who determine where scientific institutions are to be located, for they do not take into consideration the difficulties faced by scientists due to uneven distribution of information. It seems to us that in determining the geographical site of a scientific center it is necessary to take into account the place it will occupy on the graph in Figure 3.

The graphs presented here raise some questions. First of all, it would be interesting to know in what way the function of journal distribution changes with time among scientific centers and how these changes influence the efficiency of scientific creativity, which can be measured by the number of citations or at least by the number of publications in serious journals. Still, it can be suggested that in the recent past, say around 1900, the picture was quite different: at that time there was a smaller number of scientific centers and the distribution of journals among these centers seemed to be much more even. In the example given above of Odessa University, during the nineteenth century the library

TABLE 1.   *Percentage of creative scientists, journal titles, and total journals in four centers*

City	Creative scientists	Journal titles	Total journals
Moscow	50	18.9	29.2
Leningrad	13	11.8	14.7
Kiev	7	5.2	4.3
Kharkov	5	2.9	2.0

had been stocking at a rate equal to that of the growth of the number of scientific journals in the world. It remains unclear whether the irregularity of the distribution function will increase further, but there are indications that it may do so. Both the scientific centers and the number of new journals will grow, and so far this process has been going on faster then the growth of the national income.

The second question is to what degree the existing distribution of journals among scientific centers meets the needs of scientists. We lack the necessary data to give a precise answer to this question, but it follows from the materials at our disposal that Moscow receives a bit more than 10 thousand foreign scientific journals of different titles. To be more precise, this is the number of journals available to a wide range of Moscow scientific workers. [The *Cumulative Catalogue of Foreign Scientific Journals in the USSR Libraries* does not include the journals entering VINITI (The All-Union Institute for Scientific and Technical Information), for it has no public library.] The data about foreign scientific periodicals (see, for example, Nalimov and Mul'chenko, 1970) indicate that in Moscow about one-third of the total number of foreign scientific journals are available to a scientist. Taking into account the dispersion law of Breadford (Nalimov and Mul'chenko, 1969) and the fact that the most serious journals are among those available, the situation seems quite satisfactory. However, the following example demonstrates that the number of journals still sometimes proves insufficient.

When compiling a bibliography of papers which used statistical methods of experimental design, we searched the literature on the basis of *Science Citation Index*. It turned out that the Moscow University library contained about 50 percent of the publications scanned by *Science Citation Index*, and another 30 percent were found in other Moscow libraries or book depositories. The remaining 20 percent of the literature proved unavailable, even though it included some papers of great importance.

As far as provincial scientific centers are concerned, they are generally at the end of a distribution channel. This situation may sometimes be alleviated by the fact that the number of scientific specialities in these centers is not large, and thus their scientific activity can be provided for by a small number of journals. But surely this is not always the case.

There is also another problem. The editors of the journal *Industrial Laboratory* receive a large number of manuscripts that give poor solutions of successfully solved problems. Lack of information causes people to write poor papers and then others waste time reviewing them! Moreover, sometimes papers of this type do get published and cause a further flow of publications, resulting in waste of the means allocated to science. We offer no solution to the problem.

And finally, the third question: What is the distribution of journals among scientific centers in countries other than the USSR? Although we have not studied this question, we have noticed that the great concentration of scientists in a few centers characteristic of the USSR occurs in only a few European countries. In the CSSR 52 percent of the creative scientists are concentrated in Prague; in Hungary 60 percent are in Budapest, and in Denmark 62 percent are in Copenhagen. However, in Great Britain the distribution of creative scientists is as follows: London, 23 percent; Cambridge, 5 percent; Oxford, 4 percent; Birmingham, 3 percent; Glasgow, 3 percent; Manchester, 3 percent; etc. Certainly, it would be interesting to trace the way information is geographically distributed in these countries.

The data given in this section suggest the following hypothesis: the high concentration of scientists in a few centers is observed in the countries where there is an acute deficiency of the means allotted for the acquisition of journals. This lack of means leads to the high concentration of journals in a few centers, and naturally the scientists tend to be found in the centers with a high information level. The data given in the following section are further proof of this hypothesis.

Distribution of the Journals *Biometrika,* *Technometrics,* and *Lancet* Among the Countries of the World

The editorial boards of two journals, *Biometrika* and *Technometrics*, have kindly put at our disposal their data about the distribution of these journals among the countries of the world.[5] These journals were not copied in the USSR (at least until 1970). In 1970, an American journal with a similar title, *Biometrics*, was copied. However, this journal is of a quite different character; it publishes mainly statistical investigations of biological or, even narrower, genetic orientation. It is also important that the leading library of the USSR, the Lenin Library, has not received *Technometrics* since 1966, though it is received by the State Public Scien-

---

[5] We do not consider here the role of various auxiliary efforts directed at achieving wider usage of journals. We have no data for undertaking a comparative investigation. This question is considered by Overhage (1969) with respect to the United States. He deals with the arranging of a special network (information networks or networks of knowledge) for the distribution of scientific information within a country. The system consists of many elements. One of them is a new form of the arrangement of the inter-library service: the catalogue of 1968 contains 416 libraries of the United States and Canada using the teletype for inter-library service. Another element is the organization of clearing houses oriented to special themes that propagate information in a narrow field: newsletters, bulletins, bibliographies, reviews, interpreting materials. The third element is the organization of hierarchical information centers with wide usage of computers. In the field of medicine, such a center provides information service for physicians, scientists, students, librarians, etc.

Geographical Distribution of Information

tific and Technical Library. Evidently, the library considers this journal purely technical though in fact this is not the case.

Biometrika is published in Great Britain and is devoted to methodological problems of mathematical statistics; this journal was founded in 1901. The second journal is published in the United States. It also deals with methodological problems of mathematical statistics, but the papers published are more practically oriented; they attempt to solve engineering problems and those of physical and chemical investigations. *Technometrics* was founded in 1959. Both of these journals have an international character. One can hardly find a scientist who seriously studies the methodological problems of mathematical statistics without turning to these journals; they are also necessary to all those engaged in consulting services involved with the possibility of applying non-trivial statistical methods of investigation in experiments. Because new statistical methods of investigation are penetrating every field of ex-

TABLE 2. *Distribution of the statistical journals* Biometrika (B) *and* Technometrics (T) *among the countries of the world in the early 1970's*

Country	B	T	Country	B	T
Africa*	79	37	Japan	221	109
Argentina	22	6	Jordan	1	—
Australia	52	69	Kuwait	—	1
Austria	4	1	Lebanon	1	—
Belgium	10	14	Liechtenstein	1	—
Brazil	23	12	Malaysia	5	1
Canada	77	184	Mexico	10	17
Chile	9	4	Netherlands	36	43
China	13	7	New Zealand	12	11
Colombia	3	3	North and South Korea	5	4
Costa Rica	1	1	Norway	10	7
Czechoslovakia	11	4	Pakistan	13	5
Denmark	21	28	Panama	1	—
Finland	10	11	Peru	—	4
France	81	43	Philippines	5	3
Germany	98	57	Portugal	10	5
Great Britain	545	2,265	Rumania	1	1
Greece	4	2	Saudi Arabia	1	2
Guatemala	1	1	Singapore	1	3
Hungary	7	5	Sweden	23	35
Hong Kong	3	2	Switzerland	16	14
Iceland	1	1	Thailand	2	2
India	128	52	Turkey	3	1
Indonesia	—	1	United States	1,250	3,428
Iran	4	3	Uruguay	4	2
Iraq	3	2	USSR	19	20
Israel	9	15	Venezuela	8	8
Italy	64	62	Yugoslavia	6	5

* The subscriptions to *Biometrika* break down as follows among the African countries: Algeria, 1; Angola, 3; Cameroun, 1; Congo, 2; Ethiopia, 2; Ghana, 4; Ivory Coast, 1; Kenya, 1; Libya, 2; Mozambique, 1; Nigeria, 7; Republic of South Africa, 28; Senegal, 1; Sierra Leone, 1; South Rhodesia, 5; The Sudan, 2; Uganda, 2; United Arab Republic, 10; Zambia, 5.

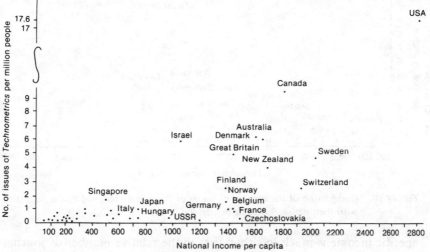

FIGURE 9. Comparison of the number of issues of *Technometrics* received by the countries with their national income per capita (in American dollars).

perimental activity, these journal must be regarded as essential to all comprehensive libraries.

Table 2 presents the data concerning the number of issues of these journals distributed among the countries of the world. The unevenness of the distribution of these journals is striking. Moreover, it does not correlate with our idea of the scientific productivity of some of the countries. For example, it seems strange that such countries as the Republic of South Africa, Israel, and the Philippines receive *Biometrika* in such great quantities; it is also subscribed to by such tiny countries as Saudi Arabia, Senegal, and Liechtenstein. At the same time, the number of subscriptions in the Soviet Union is relatively small, and *Technometrics* is not received at all by Poland.

Figures 9 and 10 relate the number of journals per million of population with the income[6] per capita. The points on the graphs are distinctly divided into two groups: for poor countries (with per capita income lower than $1,000) no correlation is observed, whereas for the rich countries one can see a marked connection between the specific income and the relative number of journal subscriptions. However, the variance region is very large, and some countries deviate from the general tendency. It is noteworthy that even for rich countries the difference in the

---

[6] The data concerning the income per capita are taken from Simpson (1968). We shall not give a critical analysis of these data here. Note only that their comparability with Soviet data is dubious since the latter are calculated by a different method.

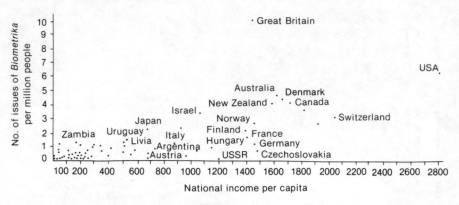

FIGURE 10.   Comparison of the number of issues of *Biometrika* received by the countries
with their national income per capita.

specific income is much less than that in the relative number of journal
subscriptions. If, say, we take the countries that fall into a threefold interval with regard to income (from $1,000 to $3,000 per capita), then the
number of journals in these countries per million of population will
change by two orders of magnitude.

Price (1967, 1970) insists on comparing indices typical of science "size"
with a summarized gross national product of countries[7] rather than with
their comparative per capita income. Using Price's data we compared the
number of the journals received with the total gross national product of
countries, the number of creative scientific workers, and the number of
articles published in the fields of physics and chemistry. These data,
presented as percentages in Table 3, show that in some countries the
number of publications in physics and chemistry correlates perfectly with
the gross national product.[8] In the first case $V = 0.79$; in the second,
0.86. The number of creative scientists is also connected with the total
national income and the number of publications.[9] Nevertheless, one can
trace considerable departures from this regularity (e.g., in Great Britain).
In some countries scientists are provided with more favorable conditions
than the average and in other countries with much worse conditions.
Figure 11 demonstrates the distribution of countries according to the
number of journal subscriptions per scientist. These graphs serve as a

[7] Gross national product (GNP) differs from national income only by taking into consideration the
amortization on credit.

[8] Certain violations can be explained easily; e.g., the large number of publications for The Netherlands
is accounted for by the fact that this country publishes a variety of international journals.

[9] The data as to the number of creative scientists in the USSR are considerably diminished here. Price
took them from Garfield's *International Directory*, compiled on the basis of *Current Contents*, which
scans the journals according to their citation. A previous paper already stressed rather low citation of the
Soviet papers in foreign publications and accounted for it (Nalimov and Mul'chenko, 1969).

measure of differences in the scientific information available to scientists in the field of mathematical statistics throughout the world.

More recently, we have obtained data on the world distribution of subscriptions to the well-known medical journal *Lancet*. These data, given in Table 4, can be compared with the similar data for *Biometrika* and *Technometrics*. In the case of *Lancet*, the distribution is even more uneven. This might result from the fact that *Lancet* is intended not only for scientific workers but also for physicians, and this raises the question of how information is distributed among those using scientific results in practice.

Concluding this section, we again formulate two questions: Is such uneven distribution of information characteristic of other branches of knowledge, and in what way might it influence further scientific progress?

TABLE 3. *Comparison of scientific activity and gross national product (GNP) of various countries* *

| | | Percentage of total | | | | |
| | | Publications in: | | | Subscriptions to: | |
Country	GNP	Physics	Chem-istry	Scien-tists	*Biomet-rika*	*Techno-metrics*
United States	32.8	31.6	28.5	41	41.8	73.7
USSR	15.6	15.6	20.7	8.3	0.64	0.6
Federal Republic of Germany	5.2	6.2	6.3	6.7	3.3	1.2
Great Britain	4.8	13.6	6.7	16.4	18.2	5.7
France	4.5	6.3	4.5	5.4	2.7	0.92
Japan	3.6	7.8	7.3	4.1	7.4	2.3
Italy	2.6	3.4	2.7	2.2	2.1	1.3
Canada	2.2	1.1	2.0	3.2	2.6	4.0
India	2.2	1.8	2.2	2.3	4.3	1.1
Poland	1.6	1.5	2.9	1.0	0.07	0
Australia	1.1	0.5	1.2	1.6	1.7	1.5
Rumania	1.0	0.6	0.9	0.4	0.03	0.02
Spain	0.9	0.2	0.4	0.2	0.74	0.37
Sweden	0.9	0.7	0.9	1.3	0.77	0.75
Netherlands	0.9	5.2	0.8	1.1	1.2	0.92
Belgium	0.8	0.3	0.6	0.7	0.53	0.3
Czechoslovakia	0.7	0.9	1.6	1.4	0.37	0.09
Switzerland	0.7	1.0	1.0	1.4	0.53	0.3
Hungary	0.5	0.5	1.0	0.8	0.23	0.1
Austria	0.4	0.2	0.5	0.5	1.3	0.02
Bulgaria	0.4	0.2	0.5	0.3	1.3	0.02

* The first three columns are taken from Price (1967) and the fourth from Price (1970).

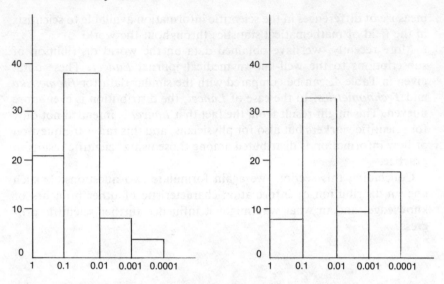

FIGURE 11.   Distribution of countries according to the number of issues of *Biometrika* (left) and *Technometrics* (right) received per active scientific worker. On the abscissas are plotted the intervals for the relation of the number of issues to the number of active scientific workers; on the ordinates, the number of countries.

World Distribution of *Science Citation Index*

Garfield's *Science Citation Index* (*SCI*) is an excellent, though very expensive, source for a rapid search of scientific information. It is also a unique source for watching the development of science (Nalimov and Mul'chenko, 1969). We analyzed the distribution of subscriptions to *SCI* distributed among the countries of the world in 1971.

Again, we noted the manifestation of extreme unevenness. *SCI* was distributed to only 36 countries. The United States obtained 422 copies, such a small country as The Netherlands received 11, and the USSR received only 3. Obviously, scientists of different countries work under extremely unequal conditions as far as their access to information is concerned.

Concluding Remarks

Though of a fragmentary character, this work is aimed at asking questions rather than answering them. We shall regard our task as fulfilled if

TABLE 4. *Distribution of the journal* Lancet *in 1975**

Country	No. of sub-scriptions	Country	No. of sub-scriptions
Afghanistan	5	Lebanon	64
Algiers	5	Liberia	5
Angola	7	Libya	11
Antilles (West Indies)	15	Luxembourg	6
ARE	71	Malawi	8
Argentina	156	Malaysia	87
Australia	1,332	Malta	13
Austria	56	Mauritius	8
Bahamas	9	Morocco	5
Barbados	8	Mozambique	7
Belgium	415	New Zealand	294
Bolivia	8	Nepal	10
Brazil	181	Netherlands	1,710
Bulgaria	6	Nicaragua	5
Burma	28	Nigeria	81
Canada	2,452	Norway	217
Chile	27	Pakistan	91
China (and Taiwan)	6	Panama	5
Colombia	37	Papua New Guinea	10
Costa Rica	8	Paraguay	7
Czechoslovakia	33	Peru	26
Denmark	659	Philippines	33
Ecuador	14	Poland	93
Ethiopia	18	Portugal	1,152
Federal Republic of Germany	546	Puerto Rico	4
Finland	306	Rumania	23
France	441	Saudi Arabia	15
Gambia	3	Sierra Leone	4
German Democratic Republic	2	Singapore	55
Ghana	53	Somalia	3
Gilbraltar	3	South African Republic	749
Great Britain	12,311	South Korea	39
Greece	344	Southern Rhodesia	43
Greenland	11	South Vietnam	7
Guatemala	13	Sudan	9
Guyana	12	Swaziland	6
Honduras	2	Sweden	952
Honduras, British	2	Switzerland	424
Hong Kong	65	Syria	11
Iceland	374	Taiwan	23
India	665	Tanzania	26
Iran	55	Thailand	48
Iraq	37	Trinidad and Tobago	46
Israel	506	Tunisia	7
Island of the Pacific Ocean	18	Uganda	24
Italy	1,005	United States	19,838
Jamaica	46	Uruguay	11
Japan	955	USSR	12
Jordan	13	Yemen	7
Kenya	44	Yugoslavia	111
Kuwait	15	Others	144

* The percentage distribution by areas of the world was as follows: North America, 44.5%; Great Britain, 24.7%; Western Europe, 16.1%; Far East, 4.3%; Australia, 3.3%; Africa, 2.3%; South and Central America, 2.0%; Middle East, 1.9%; Eastern Europe, 0.9%.

this paper gives an impetus to further studies in this direction. In any case, at present sciencemetrics states a new problem: that of unequal provision of information to scientists. The sphere of scientific activity is being entered by many new countries, some of which are rather poor. If the difference in material resources does not decrease with time, an increase in the inequality of access to information is inevitable. The situation worsens both as a result of the exponential growth of scientific information itself and of the increase in journal costs. The cost of scientific journals has increased so rapidly that it cannot be attributed to the inflation currently occurring all over the world. The costs of serial publications of data were 554 percent higher in 1970 as compared with the late 1950's. Within the same period, chemical and physical periodicals cost 233 percent more and those for engineering sciences increased 124 percent (Bolts, 1975). The cost of providing access to all information published grows at a faster rate than the national income.

The data presented here are but a partial illustration of this problem. If one accepts the previously developed view (Nalimov and Mul'chenko, 1969) of science as a self-organizing system, the serious character of this problem becomes obvious.

Chapter 12

On the Stock Exchange of Science

Changing Demand for Intellectuals[1]

Introduction

In what way will the social demands for intellectuals change in the near future? Whom should higher educational institutions of our country graduate? Should our educational system be modified? These are questions which worry many people, and answering them is not so easy as it might seem. We believe that observing the changes in the demand for intellectuals will allow us, if not to answer the questions, then at least to sharpen the insight of those responsible for these problems and their solution.

In some Western countries, a close watch is maintained over the rise and fall of demands for technical manpower. In the United States there is a special firm, Deutsch, Shea, & Evans, Inc., which regularly issues an engineer and scientist demand index by scanning 25 newspapers and 14 technical journals. Figure 1 shows the changes in demand for technical manpower from January 1961 to December 1972. The first thing that catches the eye is the clear non-regularity of the curve. The graph seems to support to some degree the statement made by Derek de Solla Price that the transfer from exponential growth to logistic growth will be accompanied by sharp fluctuations, marking the end to the even curve of growth. Such notable low-frequency fluctuation makes even short-term forecasting difficult. (It is hardly pertinent to speak here of any stable low-frequency harmonic constituent). Most probably, we are dealing here with a stochastic process with a temporally unstable spectrum. Statistical analysis proves useless here.

The firm Deutsch, Shea, & Evans makes only observations of the total

[1] This chapter was written with G. A. Batulova and A. V. Yarkho. A brief version was published in Russian in the journal *Priroda* (No. 2, 1975). It was translated by A. V. Yarkho.

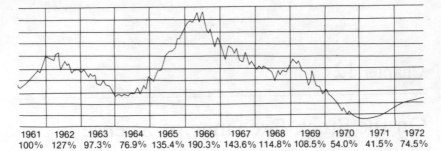

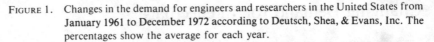

1961	1962	1963	1964	1965	1966	1967	1968	1969	1970	1971	1972
100%	127%	97.3%	76.9%	135.4%	190.3%	143.6%	114.8%	108.5%	54.0%	41.5%	74.5%

FIGURE 1.   Changes in the demand for engineers and researchers in the United States from January 1961 to December 1972 according to Deutsch, Shea, & Evans, Inc. The percentages show the average for each year.

demand, but we are interested in some other things, namely, in a detailed study of how demands for various specialities are redistributed over the course of time. These data may demonstrate the way in which the advance guard of the intellectual activities of society changes.

Such a study is more reasonably carried out by scanning employment advertisements in some country other than the United States. This is true because American professors tend to "take care" of their students by recommending them for a job. In contrast, in Great Britain the vacancies are mainly applied for by graduates on the basis of want ads.

We have analyzed the employment advertisements in two journals, *Nature* and *New Scientist and Science Journal*, for 1960 and 1969 and those in one newspaper, the *Times* of London, for 1961 and 1971 (the interval for the journals was 9 years since we began our analysis in 1970 with the journals for 1969, whereas advertisements in their present-day form only began to be published in 1960). We scanned all the advertisements in the journals and in the newspaper for January, July, and December of the two years because these are the months of the highest, lowest, and middle rate of demand. The analysis consisted in registering the number of people with various qualifications sought by the advertisers. Occasionally, there was difficulty in defining the precise number of people wanted. Sometimes the advertisements indicated the number of positions open, in which case it was registered, but some organizations demanded specialists without concrete mention of their number and in these cases we put it down as one specialist. In the majority of cases, however, it was clear how many specialists were needed.

Another approach to the analysis of employment advertisements is to record the amount of space occupied by the advertisements. The changes in this value seem to be a good index, but we gave up this approach because we found it too time-consuming. We only mention that large advertisements are published only by big firms; higher educational institutions, as a rule, publish small advertisements.

A few words should be said about statistical confidence of the results obtained. At the first stage of the research, while studying the distribution of advertisements for the principal specialties, we dealt with large samples and paid attention only to the specialties demanded comparatively often. In all these cases the relative frequencies estimated had a coefficient of variation about 5 percent, although somethings it reached 10 percent. All estimates are sure to deteriorate when we come down to analyzing the way separate small groups of specialists are distributed among institutions.

TABLE 1.   *Demand for specialists in 1960 and 1969 according to the journals* Nature *and* New Scientist and Science Journal (NSSJ)

Demand for specialists in	1960		1969	
	Nature	*NSSJ*	*Nature*	*NSSJ*
Biology	519	414	1,150	987
	(12.3%)	(9.6%)	(29.8%)	(26.4%)
Biochemistry	243	114	439	403
	(5.7%)	(2.6%)	(11.3%)	(10.8%)
Chemistry	909	984	556	554
	(21.5%)	(22.7%)	(14.5%)	(14.9%)
Physics	902	1,155	518	572
	(21.3%)	(26.6%)	(13.4%)	(15.3%)
Mathematics	443	238	232	395
	(10.4%)	(5.5%)	(16.4%)	(10.6%)
Mechanical engineering	388	949	184	151
	(9.1%)	(21.9%)	(4.7%)	(4.1%)
Metallurgy	132	110	24	47
	(3.2%)	(2.8%)	(0.7%)	(1.3%)
Agriculture	151	43	186	149
	(3.6%)	(0.9%)	(4.8%)	(4.0%)
Geology	86	69	152	124
	(2.0%)	(1.5%)	(3.9%)	(3.3%)
Medicine	281	59	83	80
	(6.7%)	(1.3%)	(2.1%)	(2.1%)
Pharmacology	79	50	130	122
	(1.9%)	(1.1%)	(3.4%)	(3.3%)
Food industry	23	28	35	24
	(0.6%)	(0.6%)	(0.8%)	(0.7%)
Geography	19	33	52	35
	(0.5%)	(0.7%)	(1.3%)	(0.9%)
Humanities	52	96	114	86
	(1.2%)	(2.2%)	(2.9%)	(2.3%)
Total	4,227	4,342	3,855	3,729
	(100%)	(100%)	(100%)	(100%)

Analysis of Employment Advertisements in
Nature and *New Scientist and Science Journal*

The results of our analysis of the two English journals are given in Table 1. The data support the following conclusions:

1. The total number of employment advertisements hardly changed during the 9-year interval. Note that Figure 1 shows roughly the same picture for the United States. A kind of synchronous development of these two Western countries seems to have occurred.

2. Both journals give figures demonstrating a sharp redistribution of various specialities. Biologists are required twice as often in 1969 as in 1960. Biochemists are requested twice as much according to one journal and four times as much according to the other. At the same time, the demand for physicists and chemists decreased sharply. In 1960 requests for physicists and chemists reached 43 to 49 percent, and in 1969 they were only about 28 to 30 percent. The demand for biologists and biochemists increased from 18 percent in one journal and 12.2 percent in the other

TABLE 2.   *Number of positions for biologists and biochemists advertised in the journals* Nature *and* New Scientist and Science Journal (NSSJ)

Year	Journal	Total no.	C-U	No. of positions advertised*				
				Research and development organizations				
				L-RI	Govt.	Firms	Hosp.	Total
Biologists								
1960	*Nature*	500	276	44	116	45	19	224
		(100)	(55.2)	(8.8)	(23.2)	(9.0)	(3.6)	(44.8)
	NSSJ	411	181	47	111	55	17	230
		(100)	(44.0)	(11.4)	(27.0)	(13.5)	(4.1)	(56.0)
1969	*Nature*	1,131	689	155	167	79	41	442
		(100)	(60.9)	(13.7)	(14.7)	(7.0)	(3.6)	(39.1)
	NSSJ	987	578	187	89	75	38	389
		(100)	(59.8)	(19.3)	(9.2)	(8.8)	(8.8)	(40.2)
Biochemists								
1960	*Nature*	237	105	18	12	42	60	132
		(100)	(44.3)	(7.6)	(5.1)	(17.7)	(25.3)	(55.7)
	NSSJ	113	29	8	19	19	38	84
		(100)	(25.7)	(7.1)	(16.8)	(16.8)	(33.6)	(74.3)
1969	*Nature*	429	240	39	50	43	57	189
		(100)	(55.9)	(9.1)	(11.7)	(10.0)	(13.3)	(44.1)
	NSSJ	393	216	65	25	39	48	177
		(100)	(54.9)	(16.6)	(6.4)	(9.9)	(12.2)	(45.1)

* The percentage of the total represented by each number is shown in parentheses. C-U, colleges and universities; L-RI, laboratories and research institutes; Govt., government organizations; Hosp., hospitals.

one in 1960 to 41.1 percent and 37.2 percent, respectively, in 1969. During these 9 years, science obviously turned from a predominantly physical–chemical orientation to a primarily biological one.

3. Detailed analysis of the requests for biologists and biochemists showed that these specialties, too, underwent modification (see Table 2). At present we can speak of a biologist–engineer, or of engineering biology. This follows from the fact, reflected in Table 2, that in 1969 about 40 percent of requests for biologists were made by government institutions, research institutes, and private firms. It appears that the transition of ecological problems from purely theoretical ones into real problems affecting the national economy caused industrial firms and governmental institutions to turn to biologists for help. Furthermore, a larger number of biologists were needed by firms producing foodstuffs and products of microbiological synthesis (vitamins, antibiotics, and enzymes). The activity of such firms is already of an engineering–biological character. Also noteworthy is the increased demand for biologists by medical insti-

TABLE 3. *Number of positions for statisticians and mathematicians advertised in the journals* Nature *and* New Scientist and Science Journal (NSSJ)

Year	Journal	Total no.	C-U	No. of positions advertised* Research and development organizations				
				L-RI	Govt.	Firms	Hosp.	Total
Statisticians								
1960	*Nature*	103	43	10	23	21	6	60
		(100)	(41.8)	(9.7)	(22.3)	(20.4)	(5.8)	(58.2)
	NSSJ	59	15	3	16	24	1	44
		(100)	(25.4)	(5.1)	(27.1)	(40.7)	(1.7)	(74.6)
1969	*Nature*	237	147	34	41	8	7	90
		(100)	(62.1)	(14.3)	(17.3)	(3.4)	(2.9)	(37.9)
	NSSJ	121	30	31	40	20	0	91
		(100)	(24.8)	(25.6)	(22.1)	(16.5)		(75.2)
Mathematicians								
1960	*Nature*	332	254	13	25	38	2	78
		(100)	(23.5)	(3.9)	(7.5)	(11.4)	(0.6)	(76.5)
	NSSJ	178	98	4	18	58	0	80
		(100)	(55.1)	(2.2)	(10.1)	(32.6)		(44.9)
1969	*Nature*	174†	142	9	23	0	0	32
		(100)	(81.5)	(5.2)	(13.3)			(18.5)
	NSSJ	251	127	42	43	43	6	134
		(100)	(48.6)	(16.1)	(16.5)	(16.5)	(2.3)	(51.4)

* The percentage of the total represented by each number is shown in parentheses. C-U, colleges and universities; L-RI, laboratories and research institutes; Govt., government organizations; Hosp., hospitals.

† Includes 80 for pure mathematicians.

tutions, which seems to testify to the tendency to strengthen the theoretical, i.e., the general biological, basis of medicine. Last but not least, the demand by academic institutions for biologists remained high, which suggests that the need for teachers of biology was not yet satisfied.

4. Table 3 contains data on the changes in the number of requests for mathematicians and mathematical statisticians. The first thing to be noted is a sharp increase in the need for mathematical statisticians, quite notable in both journals. In 1969, the demand for statisticians by higher educational institutions remained high, though the two journals yielded quite different results. It is of interest that statisticians were requested not only by government institutions but also by industrial firms, laboratories, and sometimes even by hospitals. One curious advertisement for a statistician was by a firm producing turkeys ready for roasting!

As to the demand for mathematicians, the data proved difficult to interpret. Whereas *Nature* showed a sharp decline, *New Scientist and Sci-*

TABLE 4. *Number of positions for chemists and physicists advertised in the journals* Nature *and* New Scientist and Science Journal (NSSJ)

Year	Journal	Total no.	C-U	L-RI	Govt.	Firms	Hosp.	Total
Chemists								
1960	*Nature*	880	348	46	142	332	12	532
		(100)	(39.5)	(5.3)	(16.1)	(37.7)	(1.4)	(60.5)
	NSSJ	974	190	74	209	492	9	784
		(100)	(19.5)	(7.6)	(21.5)	(50.5)	(0.9)	(80.5)
1969	*Nature*	544	331	48	70	78	17	213
		(100)	(60.8)	(8.8)	(12.9)	(14.4)	(3.1)	(39.2)
	NSSJ	546	307	85	45	87	22	239
		(100)	(56.2)	(15.6)	(8.3)	(15.9)	(4.0)	(43.8)
Physicists								
1960	*Nature*	862	400	40	178	199	45	462
		(100)	(46.4)	(4.6)	(20.7)	(23.1)	(5.2)	(53.6)
	NSSJ	1,142	306	56	315	430	35	836
		(100)	(26.6)	(4.9)	(27.6)	(37.6)	(3.1)	(74.2)
1969	*Nature*	502	334	30	83	23	32	168
		(100)	(66.5)	(5.9)	(16.5)	(4.6)	(6.5)	(33.5)
	NSSJ	552	333	72	66	55	26	219
		(100)	(60.3)	(13.0)	(12.0)	(10.0)	(4.7)	(39.7)

No. of positions advertised*

Research and development organizations

* The percentage of the total represented by each number is shown in parentheses. C-U, colleges and universities; L-RI, laboratories and research institutes; Govt., government organizations; Hosp., hospitals.

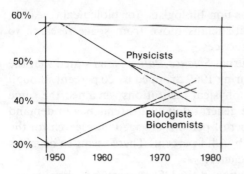

FIGURE 2. Changing proportion of physical and life scientists in U.S. scientific manpower. [Adapted from Price (1971) and Rosen (1971).]

ence Journal showed a large increase of demand. Pure mathematicians (we have selected them only from *Nature*) were required only for academic posts. Of the advertisements in *New Scientist and Science Journal*, 51.4 percent were placed by research institutes, firms, laboratories, government offices, and even hospitals. Though the last represented only 2.3 percent of the advertisements, the very fact that mathematicians were invited to work in hospitals is of great importance. We hope this is indicative of the changes beginning to take place in the thinking of physicians. The divergence of the data from *Nature* and *New Scientist and Science Journal* might relate to the fact that institutions which advertise in the latter are not purely scientific but are representative of a more administrative or technological character.

5. Despite the decline in demand for physicists and chemists, it is of interest to see where they are required. These data are given in Table 4. Both specialties remained much in demand for academic posts. In 1960, the advertisements of this kind were about 60 percent. Another interesting feature is a notable demand for both specialties in hospitals.

It would be interesting to compare the data we obtained with similar data for other countries. The graph shown in Figure 2 (reproduced from a paper by D. Price) illustrates the way the proportion of people in the United States working in physico-chemical sciences, on the one hand, and in the so-called "life sciences," on the other hand, changes. The picture is similar to that for Great Britain: at the beginning of the 1950's the respective proportions for these two fields were 60 and 30 percent, but by the end of the 1960's they were, respectively, 50 percent and 38 percent. If these curves are extrapolated, then somewhere in the 1980's the proportion to be expected is 55 percent and 45 percent. Our data for 1969 from the journals showed that chemistry occupied 28 to 30 percent of the advertisements while biology and biochemistry occupied 37 to 41 percent. Private conversations with American scientists have supported these data. They speak of the difficulties physicists and chemists face in trying to get jobs in traditional fields of knowledge and of the growing demand for biologists, as well as non-biologists, for biological research. For example, mathematical statisticians move from space research to biology, medicine, and health services.

The July 1971 issue of the journal *Science Policy News* stated that the number of jobs for physicists during 1969 dropped by 20 percent, though the number of graduates from physical institutions remained the same (14,300). Moreover, despite the general downward tendency in demand for scientific workers, the number of persons engaged in research for the Ph.D. degree in 1970 increased by 14.4 percent. There seems to be a loss of balance in the structure of science.

It is difficult to assess the situation in the USSR because we have at our disposal no data concerning the changes in demand for intellectuals. The only source of data is the reference book *USSR National Economy: 1922–1962*, which gives the number of researchers and postgraduates for the specialties in question. These figures are partly reproduced in Table 5. As the reader can see, there are no signs of the redistribution of researchers and postgraduates toward biology. The number of biologists is one-half that of physicists and mathematicians in both 1960 and 1969. Moreover, 1969 registers an increase in physicists and mathematicians. In the USSR, we picture a biologist as a teacher, sometimes as a researcher, but in no way as an engineer. All the universities except Moscow University turn out biologists on whose diplomas we read "teacher of biology."

In a 1973 issue of the Soviet journal *USA: Economy, Politics, Ideology*, we came across an article indicating that the expenditures for fundamental (purely scientific) and applied research in the U.S. universities are distributed in the following manner: 47 percent for medico-biological research, 20 percent for other natural sciences, 13 percent for technical subjects, 10 percent for socio-economic disciplines, 3 percent each for mathematics and psychology, and 4 percent for the rest.

TABLE 5.   *Percentage of researchers and postgraduate students in certain branches of science in the USSR (from the reference book* USSR National Economy *for 1970 and 1972)*

Branch of science	Researchers			Postgraduates		
	1960	1969	1971	1960	1969	1971
Physics and mathematics	8.2	10.1	10.3	9.3	11.6	10.9
Chemistry	7.4	5.0	4.7	6.5	5.5	5.2
Biology	4.3	4.1	3.9	5.1	5.1	5.1

Analysis of Employment Advertisements
in the *Times* of London

Table 6 shows that the number of requests for specialists in the *Times* in 1971 was only half as large as in 1961. Again, this is well correlated with the data for the United States (see Fig. 1). The proportion of requests for biologists among the total number of newspaper advertisements is not large, but this may be accounted for by the fact that the newspaper publishes advertisements for a wide range of specialties and many of the advertisements for biological vacancies are placed in special journals.

According to our observations, the principal newspaper advertisers are industrial firms; universities and research institutions advertise mainly in scientific journals. (However, in the above-mentioned case concerning biologists the ads were for university and college appointments.)

Another feature of interest is a mighty upward surge in demand for graduates in the humanities. The figure increased from 17.7 percent up to 25.9 percent. In 1971, this was the most frequently requested specialty; it was followed by economists (13.7 percent) and mechanical engineers (11.5 percent). Demands for other specialties were no higher than 10 percent of the total number of advertisements. The majority of the

TABLE 6. *Demand for specialists according to the* Times *of London*

Demand for specialists in	1961	1971
Biology	81 (5.1%)	94 (6.9%)
Biochemistry	23 (0.9%)	11 (0.8%)
Chemistry	181 (7.0%)	79 (5.8%)
Physics	191 (7.4%)	100 (7.4%)
Mathematics	180 (7.2%)	85 (6.3%)
Mechanical engineering	440 (17.1%)	153 (11.3%)
Metallurgy	35 (1.3%)	19 (1.4%)
Agriculture	36 (1.4%)	14 (1.0%)
Geology	26 (1.0%)	11 (0.8%)
Medicine	120 (4.7%)	53 (3.9%)
Pharmacology	15 (0.6%)	8 (0.6%)
Food industry	41 (1.6%)	2 (0.2%)
Geography	24 (0.9%)	26 (1.9%)
Humanities	456 (17.7%)	350 (25.9%)
Psychology	40 (1.6%)	30 (2.2%)
Economics	347 (13.5%)	185 (13.7%)
Management science	180 (7.0%)	104 (7.7%)
Secretaries	156 (6.0%)	30 (2.2%)
Total	2,572 (100%)	1,354 (100%)

humanists were sought by colleges (65.6 percent), the percentage of these advertisements having doubled since 1961 (32.6 percent). Demands issued by non-educational institutions were distributed among firms, government offices, and laboratories. Applications were invited from lawyers, sociologists, and graduates with a linguistic education. In many cases the last were for secretaries with a mastery of foreign languages.

In Chapter 9, the penetration of the humanities into other branches of knowledge was discussed. We have no data on this phenomenon in Great Britain, but it is evident from the U.S. university programs that great importance is attached to teaching the humanities to students in other fields. This seems to stem from the growing apprehension of the fact that humanistic education promotes the development of students' ability to think independently better than the technical sciences or even the exact ones. What society needs at present is not only narrow specialists but also researchers who bring a broad interdisciplinary background to their work.

Concluding Remarks

We believe that the type of data discussed above should be taken into consideration while determining the direction education should take. Such data should be obtained systematically and, probably, for a larger number of countries in order to provide a clearer picture of how demands for specialists change.

It is probably pertinent to make use of other data as well, such as salary changes for various specialties. At present, we have only scattered information. For example, in 1971 *Science Policy News* mentioned that the median salary of a mathematical statistician in the United States was one of the highest in the country: 16,000 dollars per year.

We think demand analysis will prove useful in forecasting almost any prospects. For example, if we wish to imagine in what direction and with what rate the equipment should change in higher educational institutions, it might be helpful to consider curves of parameter growth constructed for various devices advertised in journals.

Chapter 13

Instead of a Conclusion[1]

The theory of cognition is
the philosophy of psychology.
LUDWIG WITTGENSTEIN

The year 1981 marks the 200th anniversary of Kant's work *The Critique of Pure Reason*. This book and subsequent critical papers by Kant formed an entire epoch in European philosophical thinking. The key question of science seems to have been formulated: In what way can knowledge be achieved? But scientific development went its own way independent of Kantian criticism. Naive optimism was generated by the process of scientific development, by its tangible achievements in mastering the world.

Critical or, if you like, epistemological consideration of the scientific method of understanding the world started comparatively recently. The first impetus to this was the troubles with the foundations of mathematics. Later, criticism spread to other branches of knowledge.

In discussing anew the problem formulated by Kant, we rely upon the variety of experience accumulated by scientific development in the post-Kantian period. It is noteworthy that modern statements are surprisingly in accord with Kantian ideas. Despite all the successes of the natural sciences, we still confront the impossibility of constructing scientific knowledge on a purely empirical base. At the same time, the many successful deductive constructions in science do not make it permissible to ascribe the crucial role entirely to the logically acting mind. In my view the key problem formulated by Kant can be stated as a question: In what way does our scientific knowledge depend upon our antecedent ideas? I would like to dwell upon this problem here.

---

[1] Translated by A. V. Yarkho.

271

Any deductive conception outside mathematics is built arbitrarily and vaguely by use of polysemantic words. This may be interpreted as the presentation of initial premises by means of a fuzzy field of judgments for which the distribution of probabilities is given. Any scientist, before he becomes acquainted with a new conception, has his own system of prior notions which also have a distribution of probabilities. The construction of new concepts takes place on this field of statements probabilistically weighted in a new way: thus, the initial statements are logically developed, often with the support of new, empirically obtained facts. Here again, everything has a probabilistic flavor, as was shown in Chapter 2 of this book. It becomes clear why every serious trend of thought has many ramifications, though each of them seems to be based on the same initial system of statements and to reach its conclusion by means of the same Aristotelian logic. Among philosophers, there are several schools of thought on Kantian philosophy, and there are various interpretations of the teachings of Hegel. Similarly, diverse groups consider themselves representatives of Marxism, and Darwin's theory has also been interpreted in several ways. One of the interpretations of Darwin's theory, the only one acknowledged for a period of time in the USSR, was that connected with the name T. D. Lysenko. (After an end was put to his influence, molecular biology began to flourish in the USSR; Soviet researchers can now boast of certain achievements in practical genetics.)

Mathematical statistics includes several ramifications of "Bayesian statistics." Any natural–philosophical interpretation of a particular mathematical result is always arbitrary: we always rely upon the system of our prior notions. The variety of opinions without which we cannot imagine our intellectual life can be regarded as a result of the weighing of one system of assumptions against another one, given a priori. What is the source of the prior system of assumptions? On the one hand, it is given by education, past experience, and upbringing; on the other hand, it reflects inborn psychic peculiarities. Is not the conflict between scientific opinions reduced sometimes to the struggle of various genotypes? [On the influence of different genotypes on behavior, see, for example, Davis (1975).]

Such an interpretation of the role of prior concepts is, of course, essentially different from the Kantian one; nevertheless, one can observe here an echoing of ideas.[2] Is not Wittgenstein's famous proposal that the

---

[2] Much earlier the role of apperception was formulated in the words: "He that hath ears to hear, let him hear." However, Christ was heard differently by different people, and Christianity, during its twenty centuries of existence, spread in many directions. Some followers took up arms, and we heard the ringing of swords. Others took note of the opening words of the Sermon on the Mount: "Blessed are the poor in spirit: for theirs is the Kingdom of Heaven." Should we attribute these differences to a lack in some people of the selective prior distribution function for initial concepts which inevitably distorts their perception of any teaching?

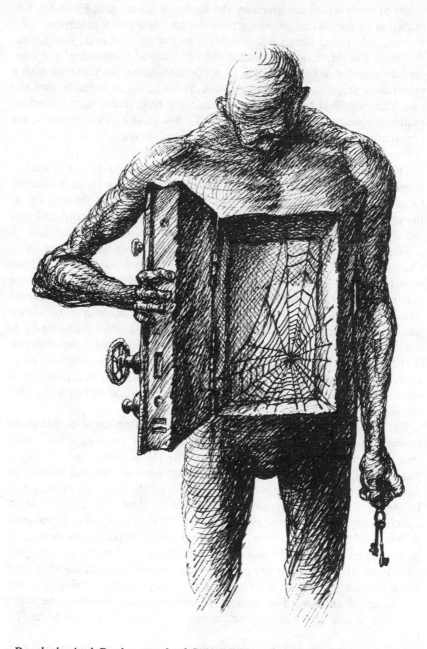

*Psychological Background of Science: In what way does our scientific
knowledge depend upon our prior store of ideas?*

limits of one's world are given by the limits of his language just a reformulation of Kant's insight concerning the role of a priori categories? We
can regard space, time, and causality as purely linguistic categories.
Probably the reluctance to discuss the ontological meaning of the concept "chance" can be explained by a correlation of this concept with a
priori categories, in the Kantian sense. We might also suppose that the
role Kant ascribed to mathematics has been resurrected in the modern
tendency toward the mathematization of knowledge and has found a new
explanation in interpreting mathematics as a language.

This book consists of separate fragments. Each chapter is a glance at
science from a very specific angle. I have arbitrarily rejected any attempt
to build an all-embracing system. All we can afford is a nibbling away at
the problem. Humanity has created a system whose complexity and versatility do not yield to any all-embracing description.

Some readers will probably note that, from all angles, science is criticized. This book is, to a large degree, conceived as a dialogue with those
who regard the cognitive power of science with naive optimism. As a
matter of fact, this optimism has started to weaken, and critical papers
have begun to appear. For example, one Soviet mathematician with a
philosophical disposition (Shreider, 1969) suggested that science serves as
a source for both knowledge and superstitions. However, the criticism
that has appeared does not yet allow us to consider the problem from
every angle. And we are left at the end of the book without an answer to
the awe-inspiring question of whether the World could be cognized with
the help of science.

But perhaps the above question is not correct. We must realize that the
assertive part of this question proceeds from the unconditional belief
that the idea of cognizability is meaningful. The paradigm of the recent
past was quite certain on this point: if it was obvious that the world was
constituted of elementary blocks structured by the Laws of Nature and
Universal Constants that were unchanging in time, the idea of world
.cognizability was also obvious, and the question as to whether it was possible could be asked. But now, when the previous initial premises start to
be perceived as naive, the question itself should be removed from consideration. It would be more reasonable to replace it with another one:
What has science with its urge toward world cognition contributed to the
evolution of culture? The answer to this question seems to be more
meaningful and at the same time more stimulating. It will sound as follows:

1. The urge toward World cognition has led to more and more profound *mastery* of the World (see Chapter 1). This can be profusely illus-

trated. One example is the ecological crisis which signifies, so to speak, the *over-mastery* of the World.

2. Actually, scientific theories do not explain anything since they always make us face new, more serious problems. We would rather say that they *reconcile* us to the new knowledge discovered in the infinite process of the mastery of the World. We say that they reconcile us because within these theories new phenomena are arranged according to the prior categories familiar to us, which had been distinctly seen by Kant. And if we dare to erect a bridge between Kant and Jung (at a very deep level their conceptions come very close to each other: they both — one of them at a logical level and the other at a psychological one — attempted to discover what had been introduced into our consciousness by the mere fact of our being Homo sapiens), we shall have to acknowledge that symbolic images, archetypes of our unconscious, playing the part of the *structuring* operators in the construction of new theories, are psychologically the factors which reconcile us (at the unconscious level) with our constructions.[3]

3. Perhaps the most remarkable difference between science and all other fields is that in the process of its evolution science *perpetually* makes its participants face new, more and more complicated problems which, in their turn, allow a glimpse at the new mysteries of the World. The development of science is a growing dramaturgy of discovering new, previously unseen mysteries. Earlier, mankind faced new mysteries only with the advent of great teachers, such as Buddha or Christ, but now the discovery of new mysteries has become almost an everyday occurrence. The intellectual peace of the World is broken.

4. We have to acknowledge with a certain amazement that the depths of our unconscious are remarkably bottomless. Our unconscious seems to be able to generate mutually exclusive conceptions of the World and to reconcile them with one another. For almost three centuries, Western science has been living in the clutches of rigid determinism, which was pressed on us by Judaism through the Old Testament and thus was at the source of our culture. The same unconditional determinism can be found in the psychology of peoples who are now at the stage of primitive culture, and this seems to form a natural conclusion that this idea is typical

---

[3] I am influenced here by the article "Der Einfluss archetypischer Vorstellungen auf die Bildung naturwissenschaftlicher Theorien bei Kepler" (Pauli, 1952). It is a pity that studies of this sort were not continued. As a matter of fact this paper has much in common with the work of Hadamard (1949) devoted to the study of psychological foundations of the creative process of mathematicians. The thrust of Hadamard's book can be formulated as follows: neither mathematical symbols nor words of our everyday language participate in the process of mathematical creativity. Rather, it takes place at the unconscious level. And it may be that this fact reconciles us with new concepts to a greater degree than the accompanying formal proof.

of the unconscious. But now some branches of contemporary science, especially of physics, are inclined to reject the idea of Fundamental Laws and Universal Constants and propose instead the idea of the spontaneous development of the World (see, for example, the "bootstrap" concept of Chew, 1968) or of the invisible holisticity of the World as a hologram, when the laws of the whole are not the goal of the research but a motion toward the permanently emerging "new integrity" (Bohm, 1973). Of great interest are the papers of the American physicist Capra: one of them contains a comparison of the ideological premises of modern physics and Oriental mysticism (Capra, 1976), and in another paper physical ideas are juxtaposed with Tao (Capra, 1975). In an earlier publication, I attempted to show that predecessors of cybernetics can be discovered in Ancient Indian philosophy. All this, perhaps, testifies to the fact that science in its progress scoops from the storerooms from which something has already been scooped. And if this is so, science becomes the Demiurge who reveals the variety of possible visions of the World within the boundaries of one culture. Historically, open polymorphism (i.e., not driven to the sectarian underground) in the comprehension of the World is also a sufficiently rare phenomenon.

5. The acknowledgment of the right to describe a phenomenon by a number of inconsistent models, the awareness of the fact that chance is in no way the expression of our ignorance but, on the contrary, one of the ways to present knowledge, and the use of semantically fuzzy concepts to enrich our scientific language (Nalimov, 1981) all broaden the limits of our consciousness generated by science.

6. In the process of its evolution, science not only destroys fundamental conceptual structures of our culture. It does something more: it also destroys the categories of our consciousness through which it arranges our sensory perception. The category of causality in its hard understanding, after millenia of stubborn resistance, sank to the background. Now the categories of time and space are being attacked. The first rather painful blows were struck by Einstein and Minkowski. More recently, the journal *Foundations of Physics* has published numerous articles on the problems of time and space in contemporary physics:

"Comments on the Dimensionality of Time" by R. Mirman (3:321–333, 1973)

"Structure of Hadron Matter: Hierarchy, Democracy or Potentiality" by A. Kantorovich (3:335–349, 1973)

"Is Time Dilation Physically Observable?" by W. Kantor (4:105–113, 1974)

"Time, the Grand Illusion" by H. G. Ellis (4:311–319, 1974)

"Tachyons and Superluminal Wave Groups" by G. H. Duffey (5:349–354, 1975)

"Extended Relativity: Mass and the Fifth Dimension" by J. D. Edmonds (5:239–249, 1975)

"Relativity and the Status of Becoming" by M. Câpek (5:607–617, 1975)

"New Four-Dimensional Symmetry" by J. P. Hsu (6:317–339, 1976)

"Interaction, Not Gravitation" by R. Schlegel (6:435–438, 1976)

"Arbitrariness of Geometry and the Aether" by P. E. Browne (6: 457–471, 1976)

"Laser Experiments and Various Four-Dimensional Symmetries" by J. P. Hsu (7:205–220, 1977)

"Projective Spacetime" by B. E. Eichinger (7:673–703, 1977)

"Remarks on Stratifying Interactions in Five Dimensions" by F. T. Vertosick (8:93–102, 1978)

"Spacetime Code: Preliminaries and Motivations" by G. McCollum (8:211–228, 1978)

"Conventionalism and General Relativity" by I. W. Roxburgh and R. K. Tavakol (8:229–236, 1978)

"Quantum Electrodynamics Within the Framework of a New Four-Dimensional Symmetry" by J. P. Hsu (8:371–391, 1978)

"The Analysis of Time: Is the Relativistic Time Unique?" by J. P. Hsu (9:55–69, 1979)

The titles of these articles are extremely interesting from a philosophical standpoint. It can be seen that the difficulties physics has to face while structuring spatial and temporal phenomena account for the multitude of inconsistent ideas about the nature of time and space. And it is impossible to choose one concept on the basis of an experiment. The physical world is no longer viewed through a set of theories, but through the multitude of fundamental notions concerning the foundations of the World, while the foundations ontologically become illusory. Perhaps contemporary physics supports Kant's idea that the categories of time and space (forms of sensory contemplation in his terminology) have a linguistic nature. And if this is not the ontology of the World but only a language for the description of the World, then the attempts to modify and deepen this language are only natural.

7. There seem to exist about 200 definitions of culture. I consider one of them of special importance for us: culture as *restriction*. Psychologically, it could not be otherwise since collective activities had to be restricted and the energy had to be transformed into an arrow aimed at a single goal. Thus, our consciousness also was limited so that it selected from the abyss of the unconscious only what was permissible in this culture. Admittedly, I have made frequent use of Bayes's theorem, and now I have to resort to it once more. Broadly, it may be interpreted here as follows: the prior distribution function $p(\mu)$ is the initial limiting filter set by the paradigm of our culture, the function $p(y|\mu)$ is a new text which

represents the manifestation of free will; the posterior distribution function $p(\mu|y)$ immediately becomes a new filter. What we mean by the development of culture is the evolution of filters determined by the sequence of transitions of the type

$$p_0(\mu) \rightarrow p_0(\mu|y) \rightarrow p_1(\mu) \rightarrow p_1(\mu|y). \ldots$$

That is, the posterior distribution function at the next step becomes prior and turns into a new, slightly less rigid filter which allows us to scoop from the depths of the unconscious much more than before. Such an approach allows us to see the evolution of culture not so much as the evolution of texts but rather as the destruction of previous filters and the expansion of their carrying capacity. Here we involuntarily remember the ancient saying: "The destroying spirit is the creating spirit." The role of science seems to be destructive. Perhaps these words are a reformulation of Popper's assertion (see Chapter 1) that knowledge in science is not accumulated; simultaneously, the same saying contains the concept of the change of paradigms. However, such a description of the way science affects the evolution of culture immediately raises many questions:

How far can the expansion of filters proceed?

Is culture as a whole ready for this? In the remote past the expansion of filters took place esoterically.

Will the emergence of new and extremely complicated texts not lead to the emergence of unbreakable filters? This problem was discussed in Chapter 1, where I opposed the extreme optimism of Popper concerning the ease of changing paradigms.

But if the filter does not develop a new carrying capacity favorable for the perception of new texts, the evolution of culture will stop.

However, we are already touching on a new subject: *a probabilistic model of the unconscious.* This topic will be dwelt upon in detail in a new book which I am now completing.

What is said above makes it obvious that it is more appropriate to speak not of World cognition but rather of the *deepening* of our interaction with the World, accompanied by the *expansion* of our consciousness. I would like to believe that a critically disposed scientist regards science in this way and that it is this attitude which inspires him.

References

Abel R.   Language and the electrone. *Akten des XIV. Internationalen Kongresses für Philosophie*, 2–9 September 1968, Vienna, Vienna: Herder & Co., 1969. p. 351-6.

Akhmanov A S.   *Logicheskoe uchenie Aristotelya* (Logic teaching of Aristotle). Moscow: Izd. Sotsial'no'ekonomich. Lit., 1960. 313 p.

Aleksandrov A D.   Space and time in contemporary physics in the light of Lenin's philosophical ideas. (Omel'yanovskii M E, ed.) *Fizicheskaya nauka i filosofiya.* Moscow: Nauka, 1973. p. 102-35.

Alimov Yu I.   Use of methods of mathematical statistics for equipment data processing. *Soviet Automatic Control* 7(2):16-25, 1974.

Arbib M A.   *Brains, machines, and mathematics.* New York: McGraw Hill, 1964. 152 p.

Aristotle.   *Aristotle's metaphysics.* Bloomington: Indiana Univ. Press, 1966. 498 p.

Artsimovich L A.   Physics of our time. *Nauka Segodnya* p. 140-60, 1969.

Barber B.   Resistance of scientists to scientific discovery. *Science* 134:596-602, 1961.

Bard Y.   *Nonlinear parameter estimation.* New York: Academic Press, 1974. 341 p.

Batulova G A, Nalimov V V & Yarkho A V.   On the stock-exchange of science. *Priroda* 2:76-81, 1975.

Bazhenov L B, ed.   *Printsip dopolnitel'nosti i materialisticheskaya dialektika* (The principle of complementarity and materialistic dialectics). Moscow: Nauka, 1976. 366 p.

Berg L S.   *Nomogenesis; or evolution determined by law.* Cambridge, MA: MIT Press, 1969. 477 p.

Bhattacharya K.   The dialectical method of Nāgārjuna. *Journal of Indian Philosophy* 1:217-61, 1971.

Blekhman I I, Myshkis A D & Panovko Ya G.   Plausibility and demonstrativeness of applied mathematics. *Mekhanika tverdogo tela.* Moscow: Nauka, 1967. v. 2. p. 196-202.

Blekhman I I, Myshkis A D & Panovko Ya G.   *Prikladnaya matematika: predmet, logika, osobennosti podkhodov* (Applied mathematics: subject, logic, peculiarities of approach). Kiev: Naukova Dumka, 1976. 269 p.

Blokhintsev D I.   *Printsipial'nye voprosy kvantovoi mekhaniki* (Principal problems of quantum mechanics). Moscow: Nauka, 1966. 159 p.

Blokhintsev D I.   *Kvantovaya mekhanika. Lektsii dlya molodykh uchyonykh* (Quantum mechanics. Lectures for young researchers). Dubna, 1978. 130 p.

Bogen H J.   *Knaurs Buch der Modernen Biologie.* Munich: Droemerl Knaur, 1967. 335 p.

279

Bohm D.   *Causality and chance in modern physics.*
    London: Routledge & Kegan Paul, 1957. 170 p.
Bohm D.   Quantum theory as an indication of a new order in physics. B. Implicate and
    explicate order in physical law. *Foundations of Physics* 3:139–68, 1973.
Bohr N.   Science and the unity of knowledge. *The unity of knowledge.*
    Garden City, NY: Doubleday, 1955. p. 47–62.
Bolts D.   Lancet. *Anglia* 54:94–100, 1975.
Born M.   *Natural philosophy of cause and chance.*
    Oxford, England: Clarendon Press, 1949. 215 p.
Bourbaki N.   The architecture of mathematics.
    The American Mathematical Monthly 57:221–32, 1950.
Bourbaki N.   *Les structures fondamentales de l'analyse. Elements de mathematique.*
    Paris: Hermann, 1958. v. II. 211 p.
Bourbaki N.   *Eléments d'histoire des mathematique.* Paris: Hermann, 1960. 276 p.
Box G E P.   Statistics and the environment.
    Journal of the Washington Academy of Science 64(2):52–9, 1974.
Box G E P & Jenkins G M.   *Time series analysis, forecasting and control.*
    San Francisco: Holden-Day, 1975. 575 p.
Bross I D J.   Prisoners of jargon. *Journal of Public Health* 54:918–27, 1964.
Büchner L.   *Force and matter.* New York: Eckler, 1891. 400 p.
Byrne E F.   *Probability and opinion.* The Hague: Nijhoff, 1968. 334 p.
Capra F.   *The Tao of physics.* Boulder, CO: Shambhala, 1975. 330 p.
Capra F.   Modern physics and eastern mysticism.
    The Journal of Transpersonal Psychology 8:20–39, 1976.
Chargaff E.   Preface to a grammar of biology: a hundred years of nucleic acid research.
    Science 172:637–42, 1971.
Chew G F.   "Bootstrap": a scientific idea? *Science* 161:762–5, 1968.
Cohen F S.   What is a question? *The Monist* 39:350–64, 1929.
Corey E J & Wipke W T.   Computer-assisted design of complex organic synthesis.
    Science 166:178–92, 1969.
Davis B D.   Social determinism and behavioral genetics. *Science* 189:1049, 1975.
Dobrov G M.   *Prognozirovanie nauki i tekhniki* (Forecasting the progress of science and
    technology). Moscow: Nauka, 1969. 208 p.
Dragunov B.   Thomas Aquinas. *Filosofskaya entsiklopediya.*
    Moscow: Sov'etskaya Entsiklopediya, 1970. vol. 5. p. 380–2.
Draper N & Smith H.   *Applied regression analysis.* New York: Wiley, 1966. 407 p.
Duncan S S.   The isolation of scientific discovery. *Science Studies* 4:109–34, 1974.
Engels F.   *Lyudwig Feierbakh i konets klassicheskoi nemetskoi filosofii* (Ludwig Feuer-
    bach and the end of classical German philosophy). Moscow: Politizdat, 1973. 71 p.
Federov V V.   *Theory of optimal experiments.* New York: Academic Press, 1972. 292 p.
Feigenberg I M.   *Mozg, psikhika, zdorov'e* (Brain, psyche, health).
    Moscow: Meditsina, 1972. 111 p.
Feynman P, Leighton R B & Sands M.   *The Feynman lectures on physics.*
    Reading, MA: Addison-Wesley, 1964, 3 vols.
Fine T L.   *Theories of probability. An examination of foundations.*
    New York: Academic Press, 1973. 263 p.
Fisher I.   *Bogi, Brakhmany, Lyudi. Chetyre tysyachi let induizma* (From Brahmanism to
    Hinduism). Moscow: Nauka, 1969, 414 p.
Gaines B R & Kohout L J.   The fuzzy decade: a bibliography of fuzzy systems and closely
    related topics. *International Journal of Man-Machine Studies* 9:1–68, 1977.
Garfield E.   Negative science and "The outlook for the flying machines."
    Current Contents (26):5–16, 27 June 1977.

(Reprinted in: Garfield E. *Essays of an Information Scientist*. Philadelphia: ISI Press, 1980. 3 vols.)

Gastev Yu & Finn V. Logical truth. *Filosofskaya Entsiklopediya*.
Moscow: Sovetskaya Entsiklopediya, 1964. Vol. 3. p. 230–1.

Gellner E. *Words and things. A critical account of linguistic philosophy and study in ideology*. London: Collancz, 1959. 270 p.

Gill J G. The definition of freedom. *Ethics* 82:1–20, 1971.

Ginsburg S. *The mathematical theory of context-free languages*.
New York: McGraw-Hill, 1966. 232 p.

Gnedenko B V. *Kurs teorii veroyatnostei*. (A course in probability theory).
Moscow: Nauka, 1969. 400 p.

Gontar' V G. Similation of chemical kinetics.
Zhurnal Fizicheskoi Khimii 50(8):2154–5, 1976.

Gordon M. Evaluating the evaluators. *New Scientist* 73:342–3, 1977.

Granger C W & Hatanaka M. *Spectral analysis of economic time series*.
Princeton, NJ: Princeton Univ. Press, 1964. 299 p.

Granovskii Yu V, Duzhenko L V, Lyubimova T N, Murashova T Y, Pechnikova T A, Sterlikova N V & Strakhov A B. Analysis of the informational activity of Soviet scientists, Doctors of Chemical Sciences.
Naukometricheskie Issledovaniya v Khimii, p. 80–101, 1974.

Hadamard J. *An essay on the psychology of invention in the mathematical field*.
Princeton NJ: Princeton Univ. Press, 1949. 145 p.

Hald A. *Statistical theory with engineering applications*. New York: Wiley, 1952. 664 p.

Harran D. *Communication: a logical model*. Cambridge, MA: MIT Press, 1963. 118 p.

Hegel G W F. *Wissenschaft der Logik*. Berlin: Akademie-Verlag, 1971. 2 vols.

Helvetius C A. *Essays on the mind and its several faculties*.
New York: B Franklin, (1809) 1970. 498 p.

Herring C. Distill or drown: the need for reviews.
Physics Today 21(9):27–33, September 1968.

Heyting A. *Intuitionism: an introduction. Studies in logic and foundations of mathematics*. Amsterdam: North Holland, 1956. 134 p.

Hintikka I. The question of questions. (Mitrokhin L N, ed.) *Filosofiya v sovremennom mire*. Moscow: Nauka, 1972. p. 303–62.

Holbach P H T. *Système de la nature, ou des Louis du monde physique et du monde moral*. New York: Clearwater, (1770) 1974. 2 vols.

Horton R. African traditional thought and western science. (Turnbull C M, ed.) *Africa and change*. New York: Knopf, 1975. p. 454–519.

Hume D. The philosophical works of David Hume. (Green T H & Grose T H, eds.)
New York: Intl. Pubns. Serv., 1964. 4 vols.

Hutten E H. *The language of modern physics. An introduction to the philosophy of science*. New York: MacMillan, 1956. 278 p.

Hutten E H. *The ideas of physics*. London: Oliver and Boyd, 1967. 153 p.

Huxley J. *Problems of the relative growth*. New York: Dover, (1932) 1972. 312 p.

Ivanov V K. On incorrectly formulated problems.
Matematicheskii Sbornik. Novaya Seriya. 61(103):211–23, 1963.

John F. Numerical solution of the equation of heat conduction for preceding times.
Annali Mathematica Pura ed Applicata. Ser. 4, 40:129–42, 1955.

Jung C G. *Memories, dreams, reflections*. New York: Vintage, 1965. 418 p.

Kafarov V V, Pisarenko V N & Ziyatdinov A Sh. Planirovanie eksperimenta i kinetika promyshlennykh organicheskikh reaktsii.
Moscow: Scient. Council of Cybernetics, 1977. 30 p.

Kant I. *Werke*. Berlin: Cassirer, 1912. 6 vols.

Kasanin J S, ed. *Language and thought in schizophrenia.* (Collected papers presented at the meeting of the Amer. Psychiatric Assoc. May 12, 1939).
Berkeley: Univ. Calif. Press, 1944. 133 p.

Kendall M G. Statistical inference in the light of theory of the electronic computer.
Review of the International Statistical Institute 34:1–12, 1966.

Kenny A. *The five ways. St. Thomas Aquinas proofs of God's existence.*
London: Routledge and Kegan Paul, 1969. 131 p.

Kleene S. *Introduction to metamathematics.* Amsterdam: North-Holland, 1952. 550 p.

Klein F. *Elementary mathematics from an advanced standpoint. Arithmetic. Algebra. Analysis.* New York: Dover, 1945. 274 p.

Kolmogorov A N. Probability theory. *Matematika: eyo soderzhanie, metody i znachenie.*
Moscow: AN SSR, 1956. V. II. p. 252–84.

Kolmogorov A N, Petrovskii I G & Piskunov N S. A study of the equation of diffusion accompanied by an increase in the amount of matter and its application to a biological problem. *Problemy Biomeditsinskoi Kibernetiki Voprosy Kibernetiki.* 12:3–30, 1975.

Kompfner R. Electron devices in science and technology. *IEEE Spectrum* 4:47–52, 1967.

Kondakov N I, ed. Question. *Logicheskii Slovar'.* Moscow: Nauka, 1971. p. 79.

Korostelyov A P & Malutov M B. On estimating coefficients of genotype adaptation by means of the stationary distribution estimates of genes frequencing. *Primenenie statisticheskikh metodov v zadachakh populyatsionnoi genetiki.*
Moscow: State Univ. Press, 1975. p. 45–51.

Kuhn T S. *The structure of scientific revolution.*
Chicago: Chicago Univ. Press, 1970a. 210 p.

Kuhn T S. Reflection on my critics. (Lakatos I & Musgrave A, eds.) *Criticism and the growth of knowledge. Proceedings of the International Colloquium in the Philosophy of Science,* Bedford College, 1965.
Cambridge, England: Cambridge Univ. Press, 1970b. p. 231–78.

Kuhn T S. Logic of discovery or psychology of research. (Lakatos I & Musgrave A., eds.) *Criticism and the growth of knowledge. Proceedings of the International Colloquium in the Philosophy of Science,* Bedford College, 1965.
Cambridge, England: Cambridge Univ. Press, 1970c. p. 1–24.

Lakatos I. Falsification and the methodology of scientific research programmes. (Lakatos I & Musgrave A, eds.) *Criticism and the growth of knowledge. Proceedings of the International Colloquium in the Philosophy of Science,* Bedford College, 1965.
Cambridge, England: Cambridge Univ. Press, 1970. p. 91–196.

Lakatos I & Musgrave A, eds. *Criticism and the growth of knowledge. Proceedings of the International Colloquium in the Philosophy of Science,* Bedford College, 1965.
Cambridge, England: Cambridge Univ. Press, 1970.

Lanczos C. *Applied analysis.* Englewood Cliffs, NJ: Prentice Hall, 1956. 539 p.

Langer S K. *Philosophy in a new key. A study in the symbolism of reason, rite and art.*
Cambridge, MA: Harvard Univ. Press, 1951. 313 p.

Legostayeva I L & Shiryayev A N. Minimax weights in the problems of finding the trend of stochastic process. *Teoriya Veroyatnostei i Eyo Primenenie* 16:339–45, 1971.

Leoni E. *Nostradamus: life and literature.* New York: Exposition Press, 1961. 823 p.

Leontief V. Theoretical assumptions and nonobserved facts.
American Economic Review 61:1–7, 1971.

Levins R. The strategy of model building in population biology.
American Scientist 54:421–31, 1966.

Levy-Brühl L. *Le surnatural et la nature dans la mentalité primitive.*
Paris: Librairie Félix Alcan, 1931. 526 p.

Lifshitz M. Critical remarks on modern theory of myths.
Voprosy Filosofii 10:138–152, 1973.

Livshitz F.   The solution which does not correspond to a serious problems.
Vestnik Statistiki 10:26–33, 1973.

MacArthur H.   On the relative abundance of bird species.
Proceedings of the National Academy of Sciences of the USA 43:293–5, 1957.

MacCormac E R.   Meaning, variance and metaphor.
British Journal of Philosophy 22:2, 1971.

MacKenna S.   *Plotinus. The enneads.* New York: Pantheon, 1957. 635 p.

Mahalanobis P C.   The foundations of statistics. *Dialectica* 8:95–111, 1954.

Maksimov A A, ed.   *Filosofskie voprosy sovremennoi fiziki* (Philosophical problems of modern physics). Moscow: AN SSSR, 1952. 576 p.

Maslov V N, Nabatova L V, Nalimov VV, Nuberg I N, Ovodova A V & Slobodchikova R I.   Presentation of the results of investigation of the structural defects in germanium.
Industrial Laboratory 29:1328–33, 1963.

Masterman M.   The nature of a paradigm. (Lakatos I & Musgrave A, eds.) *Criticism and the growth of knowledge. Proceedings of the International Colloquium in the Philosophy of Science,* Bedford College, 1965.
Cambridge, England: Cambridge Univ. Press, 1970. p. 59–90.

Meadows D H, Meadows D Z, Konders J & Behrens W W.   *The limits of growth. A report for the Club of Rome; project on the predicament of mankind.*
New York: New American Library, 1972. 205 p.

Meshalkin L D & Nguen H D.   A comparison of two optimization methods.
Industrial Laboratory 32:76–81, 1966.

Meyen S V. Problem of evolutional direction.
Itogi Nauki i Tekhniki. Zoologiya Pozvonochnykh 7:66–117, 1975.

Meyen S V & Nalimov V V.   Probabilistic vision of the world and probabilistic language.
Khimya i Zhizn' 6:22–27, 1978.

Meyen S V & Shreider Yu A.   Methodological aspects of classification theory.
Voprosy Filosofii. 12:67–79, 1976.

Mirman R.   Comments on the dimensionality of time.
Foundations of Physics 3:321–33, 1973.

Mitchell R, Mayer R A & Downhowet J.   An evolution of three biome programs.
Science 192:859–65, 1976.

Monod J.   *Chance and necessity. An essay on the natural philosophy of modern biology.*
New York: Random Press, 1972. 199 p.

Monod J.   On the molecular theory of evolution. (Harrié R, ed.) *Problems of scientific revolution: progress and obstacles to progress in the sciences.*
Oxford: Oxford Univ. Press, 1975. p. 11–24.

Moravcsik M J.   The crisis in particle physics.
Research Policy. 6:78–107, 1977.

Mul'chenko Z M, Granovsky Yu V & Strakhov A B.   On scientometrical characteristics of information activities of leading scientists. *Scientometrics* 1:307–25, 1979.

Nagel E & Newman J R.   *Gödel's proof.* New York: NYU Press, 1960. 118 p.

Nalimov V V.   Yeshcho raz o sravnenii sluchainogo poiska s gradientnym metodom v simpleks planirovanii (One more time on comparing random search with a gradient method in simplex-designing). *Industrial Laboratory* 32:854–8, 1966.

Nalimov V V, ed.   *Novye idei v planirovanii experimenta* (New ideas in the design of experiment). Moscow: Nauka, 1969. 334 p.

Nalimov V V.   *Theory of experiment.* Moscow: Nauka, 1971. 207 p.

Nalimov V V.   Summary of the report in the collection of papers. *Value aspects of modern natural sciences.* Obninsk, 1973. 73 p.

Nalimov V V.   Logical foundations of applied mathematics.
Synthèse 27:211–50, 1974a.

Nalimov V V.   *Veroyatnostaya model' yazyka* (Probabilistic Model of Language).
    Moscow: Nauka, 1974*b*.
Nalimov V V.   On some parallels between Bohr complementarity principle and the meta-
    phoric structure of the ordinary language. (Bazhenov L B, ed.) *Printsip dopolnitel'nosti
    i materialisticheskaya dialektika.* Moscow: Nauka, 1976*a*. p. 121–3.
Nalimov V V.   *Yazyk veroyatnostnykh predstavlenii* (Language of probalistic concepts).
    Moscow: Scient. Council of Cybernetics, AN SSSR, 1976*b*. 60 p.
Nalimov V V.   *Nepreryvnost' protiv diskretnosti v yazyke i myshlenii* (Language and think-
    ing: discontinuity vs. continuity). Tbilisi: Tbilisi Univ. Press, 1979. 83 p.
Nalimov V V.   *In the labyrinths of language: a mathematician's journey.*
    Philadelphia: ISI Press, 1981. 246 p.
Nalimov V V & Barinova Z B.   Sketches of the history of cybernetics. Predecessors of
    cybernetics in ancient India. *Darshana International* 14(2):35–72, 1974.
Nalimov V V & Chernova N A.   *Statisticheskie metody planirovaniya ekstremal'hykh
    eksperimentov.* Moscow: Nauka, 1965. 340 p.
Nalimov V V & Golikova T I.   Logical foundations of experimental design.
    Metron 29:3–58, 1971
Nalimov V V & Golikova T I.   *Logicheskie osnovaniya planirovaniya eksperimenta* (Logi-
    cal foundations of experimental design). Moscow: Metallurgiya, 1976. 127 p.
Nalimov V V, Golikova T I & Mikeshina N G.   On the practical use of the concept of
    D-optimality. *Technometrics* 12:799–812, 1970.
Nalimov V V, Kuznetsov O A & Drogalina J A.   Visualization of verbal text by means of
    group meditation.
    Bessoznatel'noe: Priroda, funktsii, metody issledovaniya 3:703–10, 1978.
Nalimov V V & Mul'chenko Z M.   *Naukometriya. Izuchenie razvitiya nauki kak informat-
    sionnogo protsessa* (Scientometrics. Study of science as an information process).
    Moscow: Nauka, 1969. 191 p.
Nalimov V V & Mul'chenko Z M.   Science and biosphere — an effort to compare two
    systems. *Priroda* 11:55–63, 1970.
Nalimov V V & Mul'chenko Z M.   On logico-linguistic analysis of the language of science.
    (Schaumyan S K, ed.) *Problemy strukturnoi lingvistiki.*
    Moscow: Nauka, 1972. p. 534–54.
Nowakovskaya M.   *Language of motivation and language of actions.*
    The Hague: Mouton 1973. 272 p.
Oldenberg H.   *Buddha: His life, His doctrine, His order.*
    New York: Intl. Pubns. Serv., 1971 (1881). 454 p.
On teaching mathematical statistics to researchers.
    Moscow: Moscow State Univ. Press, 1971.
Overhage C F J.   Information networks.
    Annual Review of Information Science and Technology 4:340–77, 1969.
Parthasarathi A.   Universities and nation building. *Quest* 52:19–26, 1967.
Parthasarathy K R.   Probability theory on the closed subspaces of a hilbert space.
    Matematika 14:102–122, 1970.
Pauli W.   Der Einfluss archetypischer Vorstellungen auf die Bildung naturwissenschaft-
    licher theorien bei Kepler. (Influence of archetypical notions on the construction of
    scientific theories by Kepler). *Naturerklarung und Psyche.*
    Zurich: Rascher, 1952. p. 109–194.
Plato.   Theaetetus. *The dialogues.* Oxford: Clarendon Press, 1953. Vol. 3.
Podgoretskii M I & Smorodinskii Ya A.   On the axiomatic structure of constructing physi-
    cal theories. *Voprosy Teorii Poznaniya* 52:19–26, 1969.
Poincaré H.   *Science and method.* New York: Dover, 1952. 288 p.
Popper K R.   Some comments on truth and growth of knowledge. (Nagel E, Suppes P,

Tarski A, eds.) *Logic, methodology and philosophy of science. Proceedings of the 1960 International Congress.*
Stanford: Stanford Univ. Press, 1962. p. 285–92.

Popper K R. *Conjectures and refutations. The growth of scientific knowledge.*
New York: Basic Books, 1963. 412 p.

Popper K R. *The logic of scientific discovery.* London: Hutchinson, 1965. 479 p.

Popper K R. Normal science and its dangers. (Lakatos I & Musgrave A, eds.) *Criticism and the growth of knowledge. Proceedings of the International Colloquium in the Philosophy of Science,* Bedford College, 1965.
Cambridge, England: Cambridge Univ. Press, 1970. p. 51–8.

Popper K R. The rationality of scientific revolutions. (Harrié R, ed.) *Problems of Scientific Revolution: progress and obstacles to progress in the sciences.*
Oxford: Oxford Univ. Press, 1972. p. 72–101.

Price D J D. *Little science, big science.* New York: Columbia Univ. Press, 1963. 118 p.

Price D J D. Nation can publish or perish.
International Science and Technology 70:84– 90, 1967.

Price D J D. Measuring the size of science.
Proceedings of the Israel Academy of Sciences and Humanities 4:98–111, 1970.

Price D J D. Is there a decline in big science countries and in big science subjects? Presented at the 13th International Congress for the History of Science.
Moscow and Leningrad, 18–28 August, 1971.

Problemy biomeditsinskoi kibernetiki. *Voprosy Kibernetiki* (Problems of biomedical cybernetics). Moscow: Soviet Radio, 1975. 203 p.

Quastler H. *The emergence of biological organization.*
New Haven: Yale Univ. Press, 1964. 83 p.

Rabinovich B L. Alchemy as a phenomenon of culture.
Priroda 9:86–97, 1973; continued in 10:83–93, 1973.

Radhakrishnan S. *Indian philosophy.* New York: Macmillan, 1962. 2 vols.

Rastrigin L A. Criteria for comparing methods of seeking an extremum.
Industrial Laboratory 32:1529–32, 1966.

Rastrigin L A. *Etot sluchainyi, sluchainyi, sluchainyi mir* (This random, random, random world). Moscow: Molodaya Gvardiya, 1969. 222 p.

Reich C A. *The greening of America.* New York: Bantam, 1974. 294 p.

Rosen R. *The optimality principles in biology.* London: Plenum Press, 1967. 198 p.

Rosen S. Science and technology approach year 2000. *New Scientist* 50:76–9, 1971.

Rosental M M, ed. *Filosofskii slovar'.* Moscow: Polit. Literatura, 1972. 495 p.

Rosental M & Yudin P, eds. *Kratkii filosofskii slovar'.* Moscow: Politizdat, 1955. 567 p.

Russell B. *Human knowledge: its scope and limits.* London: Allen & Unwin, 1961. 538 p.

Russell B. *History of western philosophy.* London: Allen & Unwin, 1962. 842 p.

Russell B & Whitehead A N. *Principia mathematica.*
Cambridge: Cambridge Univ. Press, 1910. 3 vols.

Sartre J P. *Nausea.* New York: Penguin Books, 1965. 238 p.

Schleicher A. *Die Deutsche Sprache.* Stuttgart: Cotta, 1888. 348 p.

Schmalgauzen I I. The integration of biological systems and their self-regulation.
Bulleten Moskovskogo Obshchestva Ispytatelei Prirody. Otdelenie Biologii. Moscow, 1964.

Schmalgauzen I I. *Kiberneticheskie voprosy biologii* (The cybernetic problems in biology). Novosibirsk: Nauka, 1968. 223 p.

Shaumyan S K. *Applicational grammar as a semantic theory of natural languages.*
Chicago: Univ. of Chicago Press, 1977. 184 p.

Shcheglov V N. Obtaining a Boolean model on current information of a complicated technological process. *Industrial Laboratory* 38:70–4, 1972.

Shelestov D K.   The first Russian scientific humanitarian journal. On the occasion of 170 years anniversary of the Russian Academy of Science "Statistical Journal."
Vestnik Akademii Nauk SSSR. 5:109–15, 1977.

Sheynin O B.   J. H. Lambert's work on probability.
Archive for History of Exact Sciences 7:244–56, 1971*a*.

Sheynin O B.   Newton and the classical theory of probability.
Archive for History of Exact Sciences 7:217–43, 1971*b*.

Sheynin O B.   D. Bernoulli's work on probability.
Rete, Strukturgeschichte der Naturwissenschaften 1:274–300 1972*a*.

Sheynin O B.   On the mathematical treatment of observations by L. Euler.
Archive for History of Exact Sciences 9:45–56, 1972*b*.

Sheynin O B.   Finite random sums (A historical essay).
Archive for History of Exact Sciences 9:275–305, 1973*a*.

Sheynin O B.   Mathematical treatment of astronomical observations (A historical essay).
Archive for History of Exact Sciences. 11:97–126, 1973*b*.

Sheynin O B.   R. J. Boscovich's work on probability.
Archive for History of Exact Sciences. 9:306–24, 1973*c*.

Sheynin O B.   On the prehistory of the theory of probability.
Archive for History of Exact Sciences 12:97–141, 1974.

Shishkin I B.   Institute of Psychology of the Academy of Sciences of the USSR has come into being. *Priroda* 7:111–2, 1972.

Shreider Yu A.   Science—the source of knowledge and prejudices.
Novyi Mir 45:207–26, 1969.

Shreider Yu A & Osipova M A.   Some dynamic models in information science.
Automatic Documentation and Mathematical Linguistics. 3(3):41–5, 1969.

Simpson D.   The dimension of world poverty.
Scientific American 219(5):27–35, November 1968.

Smirnov S.   From the Cromagnon man to Kepler and from Kepler to our time.
Znanie–Sila 5:43–6, 1977*a*.

Smirnov S.   From Kepler to our time. And what is further? *Znanie–Sila* 6:39–41, 1977*b*.

Spinoza B.   *Chief works.* New York: Dover, 1955. 2 vols.

Stranden A D.   *Germetism. Sokrovennaya filosofiya Egipta* (Hermetism. A mysterious philosophy of Egypt). St. Petersburg: The University, 1914. 85 p.

Styazhkin N I.   *Formirovanie matematicheskoi logiki* (Formation of Mathematical Logic). Moscow: Nauka, 1967. 507 p.

Subbotin Yu K.   The essence of value concepts of neothomism. A critical analysis.
Nauchnye Doklady Vysshei Shkoly. Filosofskie Nauki 3:81–8, 1972.

Svechnikov G A, ed.   *Sovremennyi determinizm. Zakony prirody* (Modern determinism, laws of nature). Moscow: Mysl', 1975. 527 p.

Szaniawski K.   Formal analysis of evaluative concepts.
International Social Science Journal 27:446–57, 1975.

Taylor G R.   Prediction and social change. The need for a basis in theory.
Futures 9:404–14, 1977.

Teilhard de Chardin P.   *The phenomenon of man.*
New York: Harper & Row, 1965. 320 p.

Timofeev R E & Resovskii N V.   *Kratkii ocherk teorii evolyutsii* (A brief account of the evolutionary theory). Moscow: Nauka, 1969. 407 p.

Tolman R.   *Relativity, thermodynamics, and cosmology.*
Oxford, England: Clarendon Press, 1934. 501 p.

Toynbee A.   The religious background of the present environmental studies.
International Journal of Environmental Studies 3:141–7, 1972.

Tutubalin V N.   *Teoriya veroyatnostei* (Probability theory).
Moscow: Moscow State Univ., 1972. 229 p.

Tutubalin V N.   A review of the book: Fine, 1973.
  Novye Knigi za Rubezhom 5:28–30, 1974.
Vasil'ev A F.   Theoretical foundations of the contemporary methods of quantitative
  analysis of multicomponent systems according to the absorption spectra.
  Doctoral Thesis, Moscow, 1976. 318 p.
Vavilov S I.   *The eye and the sun; about light vision and the sun.*
  Moscow: Foreign Languages Pub. House, 1955. 135 p.
Va'yasane'ya Samhita. *The Upanishads.*
  Calcutta: Advaita Ashrama, 1957. 2 vols.
Ventsel Ye S.   *Teoriya veroyatnostei* (Probability theory).
  Moscow: Fizmatgiz, 1962. 564 p.
Watt K E F.   *Ecology and resource management. A quantitative approach.*
  New York: McGraw-Hill, 1968. 450 p.
Watts D G, ed.   *The future of statistics: Proceedings.*
  New York: Academic Press, 1968. 315 p.
Weyl H.   *Philosophy of mathematics & natural science.*
  Paterson, NJ: Atheneum, (1927) 1963.
Wiener N.   *I am a mathematician.* Boston: MIT Press, 1964. 380 p.
Wigner E P.   The unreasonable effectiveness of mathematics in the natural sciences.
  Communications on Pure and Applied Mathematics 13:1–14, 1960.
Wittgenstein L.   *Tractatus logico-philosophicus.*
  Atlantic Highlands, NJ: Humanities, 1963. 114 p.
Yakhot O.   Randomness. *Filosofskaya entsiklopediya.*
  Moscow: Sovetskaya Entsiklopediya, 1970. Vol. 5. p. 33–4.
Yourgrau W.   Gödel and physical theory. *Mind* 78:77–90, 1969.
Yule J U & Kendall M G.   *An introduction to the theory of statistics.*
  London: Ch. Griffin & Co., 1950. 701 p.
Zadeh L A.   Fuzzy sets. *Information and Control* 8:338–53, 1965.
Zadeh L A.   Quantitative fuzzy semantics. *Information Sciences* 3:159–76, 1971.
Zadeh L A.   A fuzzy-set-theoretic interpretation of linguistic hedges.
  Journal of Cybernetics 2(3):4–34, July–September 1972.
Zadeh L A.   Outline of a new approach to the analysis of complex systems and decision
  processes.
  IEEE Transactions on Systems Man and Cybernetics 3:28–44, 1973*a*.
Zadeh L A.   *The concept of linguistic variable and its application to approximate reason.*
  New York: Elsevier, 1973*b*.
Zadeh L A.   Fuzzy sets as a basis for a theory of possibility.
  Fuzzy Sets and Systems 1:3–28, 1978.
Ziman J M.   Information, communication, knowledge. *Nature* 224:318–24, 1969.

Index of Names

289

Index of Subjects

293